Lipids and Renal Disease

CONTEMPORARY ISSUES IN NEPHROLOGY

VOLUME 24

Series Editor
Jay H. Stein, M.D.

Series Editors, Vols. 1-21
Barry M. Brenner, M.D.
Jay H. Stein, M.D.

Volumes Already Published

VOLUME 3
Immunologic Mechanisms of Renal Disease
Curtis B. Wilson, M.D., Guest Editor

VOLUME 4
Hormonal Function and the Kidney

VOLUME 7
Chronic Renal Failure

VOLUME 9
Nephrotic Syndrome

VOLUME 11
Divalent Ion Homeostasis

VOLUME 12
Pediatric Nephrology
Bruce M. Tune, M.D., and Stanley A. Mendoza, M.D., Guest Editors

VOLUME 13
The Heart and Renal Disease
Robert A. O'Rourke, M.D., Guest Editor

VOLUME 14
The Progressive Nature of Renal Disease
William E. Mitch, M.D., Guest Editor

VOLUME 15
Modern Techniques of Ion Transport

VOLUME 16
Body Fluid Homeostasis

VOLUME 17
Pharmacotherapy of Renal Disease and Hypertension
William M. Bennett, M.D., and David A. McCarron, M.D., Guest Editors

VOLUME 18
Immunopathology of Renal Disease
Curtis B. Wilson, M.D., Guest Editor

VOLUME 19
Renal Transplantation
Edgar L. Milford, M.D., Guest Editor

VOLUME 20
The Kidney in Diabetes Mellitus

VOLUME 21
Atrial Natriuretic Peptides

VOLUME 22
Peritoneal Dialysis
Zbylut J. Twardowski, M.D., Karl D. Nolph, M.D., F.R.C.P.S. (Glasgow), and Ramesh Khanna, M.D., Guest Editors

VOLUME 23
Hormones, Autacoids, and the Kidney
Stanley Goldfarb, M.D., and Fuad N. Ziyadeh, M.D., Guest Editors

Forthcoming Volumes in the Series

VOLUME 25
Diagnostic Techniques in Renal Disease
Robert G. Narins, M.D., Guest Editor

VOLUME 26
The Progressive Nature of Renal Disease, 2nd Ed.
William E. Mitch, M.D., Guest Editor

Lipids and Renal Disease

Guest Editor
WILLIAM F. KEANE, M.D.
Professor
Department of Medicine
University of Minnesota Medical School—Minneapolis
Director
Nephrology Research Laboratory
Department of Medicine, Division of Nephrology
Hennepin County Medical Center
Minneapolis, Minnesota

Series Editor
JAY H. STEIN, M.D.
Professor and Chairman
Department of Medicine
Dan F. Parman Chair in Medicine
University of Texas Medical School at San Antonio
San Antonio, Texas

CHURCHILL LIVINGSTONE
New York, Edinburgh, London, Melbourne, Tokyo

Library of Congress Cataloging-in-Publication Data

Lipids and renal disease / guest editor, William F. Keane.
p. cm. — (Contemporary issues in nephrology ; v. 24)
Includes bibliographical references and index.
ISBN 0-443-08779-2
1. Kidneys—Pathophysiology. 2. Lipids—Metabolism—Disorders.
3. Hypertension—Pathophysiology. I. Keane, William F.
II. Series. III. Series: Contemporary issues in nephrology ; vol. 24.
[DNLM: 1. Kidney Diseases—physiopathology. 2. Lipids—adverse effects. 3. Lipids—metabolisms. W1 CO769MR v. 24 / QY 465 L7635]
RC903.9.L56 1991
616.6'107—dc20
DNLM/DLC
for Library of Congress 91-23471
CIP

Distributed in the United Kingdom by Churchill Livingstone, Robert Stevenson House, 1–3 Baxter's Place, Leith Walk, Edinburgh EH1 3AF, and by associated companies, branches, and representatives throughout the world.

Accurate indications, adverse reactions, and dosage schedules for drugs are provided in this book, but it is possible that they may change. The reader is urged to review the package information data of the manufacturers of the medications mentioned.

The Publishers have made every effort to trace the copyright holders for borrowed material. If they have inadvertently overlooked any, they will be pleased to make the necessary arrangements at the first opportunity.

Acquisitions Editor: *Avé McCracken*
Copy Editor: *David Terry*
Production Designer: *Angela Cirnigliaro*
Production Supervisor: *Jeanine Furino*

To Stephanie, Alicia, and Elizabeth

Contributors

Kamal F. Badr, M.D.
Associate Professor, Department of Medicine, Division of Nephrology, Vanderbilt University School of Medicine; Senior Investigator, Kidney and Urologic Disease Center at Vanderbilt University, Nashville, Tennessee

Peter C. Brazy, M.D.
Associate Professor, Department of Medicine, and Head, Nephrology Section, University of Wisconsin Medical School; Chief, Renal Section, William S. Middleton Memorial VA Hospital, Madison, Wisconsin

Jonathan R. Diamond, M.D.
Associate Professor, Department of Medicine, Division of Nephrology, Pennsylvania State University College of Medicine; Attending Physician, Department of Medicine, University Hospital, Hershey, Pennsylvania

Jaap A. Joles, D.V.M., Ph.D.
Assistant Professor, Department of Nephrology and Hypertension, University Hospital, Utrecht, The Netherlands

Bertram L. Kasiske, M.D.
Associate Professor, Department of Medicine, University of Minnesota Medical School—Minneapolis; Staff Physician, Department of Medicine, Division of Nephrology, Hennepin County Medical Center, Minneapolis, Minnesota

George A. Kaysen, M.D., Ph.D.
Professor, Department of Medicine, University of California, Davis, School of Medicine, Davis, California; Chief, Departments of Medicine and Nephrology, Veterans Administration Medical Center, Martinez, California

Paul E. Klotman, M.D.
Chief, Molecular Medicine Section, Laboratory of Developmental Biology, National Institute of Dental Research, National Institutes of Health, Bethesda, Maryland

Jeffrey B. Kopp, M.D.
Staff Fellow, Molecular Medicine Section, Laboratory of Developmental Biology, National Institute of Dental Research, National Institutes of Health, Bethesda, Maryland

Connie L. Manske, M.D.
Assistant Professor, Department of Medicine, University of Minnesota Medical School—Minneapolis, Minneapolis, Minnesota

Joel Neugarten, M.D.
Associate Professor, Department of Medicine, Albert Einstein College of Medicine of Yeshiva University; Attending Physician, Department of Medicine, Renal Division, Montefiore Medical Center, Bronx, New York

Michael P. O'Donnell, Ph.D.
Staff Physiologist, Department of Medicine, Division of Nephrology, Hennepin County Medical Center, Minneapolis, Minnesota

Brad H. Rovin, M.D.
Assistant Professor, Department of Medicine, Renal Division, Ohio State University College of Medicine, Columbus, Ohio

Detlef Schlondorff, M.D.
Professor, Department of Medicine, and Director, Division of Nephrology, Albert Einstein College of Medicine of Yeshiva University, Bronx, New York

Paul G. Schmitz, M.D.
Assistant Professor, Department of Medicine, Division of Nephrology, St. Louis University School of Medicine, St. Louis, Missouri

George F. Schreiner, M.D., Ph.D.
Associate Professor, Departments of Medicine and Pathology, Renal Division, Washington University School of Medicine, St. Louis, Missouri

Dag S. Thelle, M.D.
Professor, Department of Epidemiology, Nordiska hälsovårdshögskolan, The Nordic School of Public Health, Göteborg, Sweden

Preface

> The only way in which "clear notions on the subject (fatty degeneration of the renal epithelium in Bright's Disease) can be obtained consists in examining whether the condition of fatty degeneration is a primary or a secondary one, whether it sets in as soon as the disturbance can be perceived, our whether it does not occur until some other perceptible disturbance has gone before."
>
> R. Virchow
> *Cellular Pathology*, 1860

Altered lipid metabolism frequently coexists in patients with hypertension and complicates the clinical course of patients with diabetic and nondiabetic renal diseases. While extensive epidemiologic data have supported a critical role for hypercholesterolemia in the pathogenesis of atherosclerosis, the importance of abnormal lipid metabolism in the pathophysiology of hypertension has only recently been explored. The frequent coexistence of atherosclerotic cardiac disease in patients with diabetic and nondiabetic renal disease has suggested an important linkage between renal disease and abnormal lipid metabolism. Similarly, the possibility that the dyslipidemia induced by the presence of renal disease may also participate in kidney injury has emerged during the past decade as a new concept in our understanding of the pathogenesis of progressive renal disease.

This volume of *Contemporary Issues in Nephrology* represents the first time that a textbook has been devoted to this topical area. This timely volume begins with an important epidemiologic discussion of the associations between hypertension and hyperlipidemia, and provides an overview of the effects of dietary fatty acids on blood pressure in humans. Clinical data detailing the risk factors involved in the progression of renal disease in humans in which hypertension has been demonstrated to be a pivotal factor is reviewed. The frequent abnormalities of lipid metabolism in renal transplantation patients and the cardiovascular consequences are concisely discussed. Recent advances in our understanding of the relationships between proteinuria and dyslipidemia are presented, again underscoring the potential importance of the atherogenic profile frequently seen in these patients and new therapeutic approaches that have promise. The second half of this book focuses on the renal structural and functional consequences of dyslipidemia as well as the importance of various lipid mediators in normal and abnormal renal function. The recent experimental developments that have focused our attention on the possibilities of lipids modulating progression of renal disease are presented and the effects of lipid-lowering therapeutic interventions as well as dietary lipid manipulations are discussed. The biologic implications of lipids on the immune system as well as an up-to-date discussion of interactions of lipids with critical glomerular cells provide new and important information pertaining to the pathophysiologic effects of dyslipidemia in renal disease.

Thus, these timely presentations will be of great value to a variety of subspecialists who are frequently involved in the care of patients with renal disease.

Finally, I am grateful for the opportunity to work with all the authors and editorial staff who have greatly facilitated my task in developing what I believe will be a pivotal and frequently used text.

William F. Keane, M.D.

Contents

1

Hypercholesterolemia, Hypertriglyceridemia, and Hypertension: Are They Independent?

Dag S. Thelle

INTRODUCTION

It is generally acknowledged that coronary heart disease (CHD) is a multifactorial disease (i.e., a variety of factors are involved in the development of clinical manifestation of this disease) and, further, that the combination of these factors confers a multiplicative effect on the risk.[1] The multifactorial concept implies that more than one causal factor is operative, and that these factors, at least to some extent, are independent of each other. Each of the factors listed in Table 1-1 should have a separate independent contribution to the CHD risk, to be called a *risk factor*. In this chapter, I shall re-examine the extent to which this statement is true: for example, are hypercholesterolemia, hypertriglyceridemia, and hypertension really independent?

An impressive body of evidence from epidemiologic and clinical trials in different populations links hypercholesterolemia and hypertension to the chain of events that eventually lead to overt disease.[2,3] On the assumption of a causal relationship between these risk factors and CHD, preventive and therapeutic strategies have been developed to lower total cholesterol and blood pressure (BP) levels.[1]

Table 1-1. Risk Factors for Coronary Heart Disease

Individual traits and characteristics
Age
Sex
Family history
Blood lipids
Total cholesterol
LDL cholesterol
HDL cholesterol
Fatty acids
Triglycerides
Other laboratory variables
Sedimentation rate
Hemoglobin
Uric acid
Leukocytes
Thrombosis
Coagulation
Diabetes mellitus
Blood pressure
Overweight: waist to hip ratio
Menopause
Tissue type antigen
Nonspecific ECG changes
External factors (exposure)
Probably direct causal factors
Nutrition
High saturated fat intake
High cholesterol intake
High salt intake
Cigarettes
Alcohol abuse
Physical inactivity
Coffee
Contraceptives
Tolbutamide
Phenormin
Probably indirect associations
Vegetarianism (protective)
Life events (stress)
Soft water

Abbreviations: LDL, low-density lipoprotein; HDL, high-density lipoprotein; ECG, electrocardiogram.

Several studies have demonstrated that lowering high serum cholesterol levels reduces CHD morbidity and mortality.[3]

In the case of hypertension, on the other hand, randomized, controlled trials have failed to find any clear evidence that lowering BP reduces the incidence of CHD.[4] On the contrary, there is evidence that drug treatment may increase coronary risk in certain subgroups, as observed in the Multiple Risk Factor Intervention Trial.[5] In that study, the investigators found that men with diastolic BP less than 100 mmHg who received drug treatment had a higher mortality rate than those in the usual care group, and that intensively treated hypertensive patients with nonspecific electrocardiographic changes fared worse than those receiving usual care. Similar findings were obtained in the well-performed, although small, Oslo study, in which an excess risk of dying from ischemic heart disease was demonstrated in the actively treated group.[6]

Apart from the possibility that these are random findings, there are many potential explanations for this lack of effect on CHD:

1. Hypertensive patients may have an increased level of unidentified factors implying higher CHD risk, not corrected by antihypertensive drugs.
2. There may be an optimal low BP level below which it is harmful to further decrease BP.
3. The achieved BP reductions were too small to result in a decline of CHD risk.
4. The drugs have harmful effects that are not yet fully recognized.
5. The trials were too short and did not include the appropriate population groups.

We do not yet know the reason for failure of efficient antihypertensive drug treatment to reduce the risk of CHD substantially, and to the same extent as it reduces the risk of cerebrovascular disease. There is no doubt that BP is a strong coronary risk factor in Western societies. A recent review by Collins et al. confirmed that throughout the range of the usual diastolic blood pressure (DBP), a lower DBP was always followed by a lower risk of stroke or CHD.[7] It is true that drug trials have mainly been undertaken in middle-aged men and women, and that the "exposure time" to an increased BP may be too long for a reversal of the CHD risk by a relatively short drug intervention.

Further, Collins et al. have pointed to the modest blood pressure difference between treatment and control groups in the trials.[7] The expected long-term reductions in CHD based on the reported BP reductions would then be approximately 20 to 25 percent; within the average of 2 or 3 years of treatment between randomization and events, CHD incidence would be reduced by only 10 to 15 percent. With such small reductions, none of the reported drug trials would have sufficient statistical power.

However, even with these statistical considerations in mind, we have to investigate the other possible explanations, such as BP being a marker of a metabolic atherogenic or thrombogenic state, in which case drug-induced BP reduction may only partially reduce the associated risk of CHD. In addition, drug treatment has in some cases been shown to induce potentially detrimental metabolic changes.

The possibility that increased BP might be a marker of an untoward metabolic state is supported by cross-sectional data from large-population studies.[8–10]

In the Tromsø Study, Løchen showed that hypertensive patients had higher serum cholesterol and triglyceride levels and lower high-density lipoprotein (HDL) cholesterol.[8] In another population study in northern Norway, it was demonstrated that former users of antihypertensive drugs as well as nontreated hypertensive patients had lower lipid levels than those currently on drug treatment.[9] In addition, the drug-treated subjects had higher blood glucose levels.

In a further analysis of the Tromsø material, which included more than 15,000 untreated free-living men and women aged 20 to 54 years, it was shown that total and non-HDL cholesterol levels increased significantly with increasing systolic BP or DBP in both genders.[10] Men aged 20 to 29 years had steeper regression slopes for BP by total cholesterol than did women of similar age, and the association decreased with age in men; in women, it increased with age, as seen in Figure 1-1. When the material was stratified according to the median of the body mass index (BMI), the slope was steeper for those with the higher BMI, but still highly significant for the leaner subjects. Smoking, physical activity, and alcohol consumption had little influence on the relationship. The relationship between total cholesterol and DBP is further demonstrated in Figure 1-2, which shows how the age-adjusted frequency distribution of total cholesterol is shifted to the right in individuals with DBP greater than 99 mmHg compared with those with DBP less than 70 mmHg. The cholesterol distribution for subjects with DBP of 70 to 99 mmHg had an intermediate position.

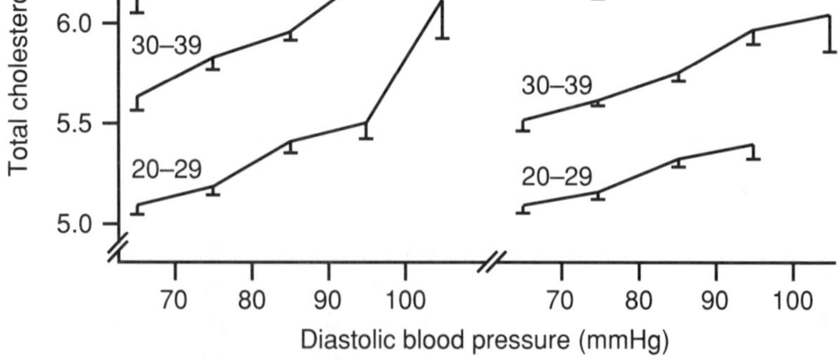

Fig. 1-1. Mean concentrations of serum total cholesterol (mmol/L) by DBP in men aged 20 to 29, 30 to 39, 40 to 49, and 50 to 54 years, and in women aged 20 to 29, 30 to 39, and 40 to 49 years. Cells with less than 20 observations were pooled with adjacent category. T bars are expressed as SE. (From Bønaa and Thelle,[10] with permission.)

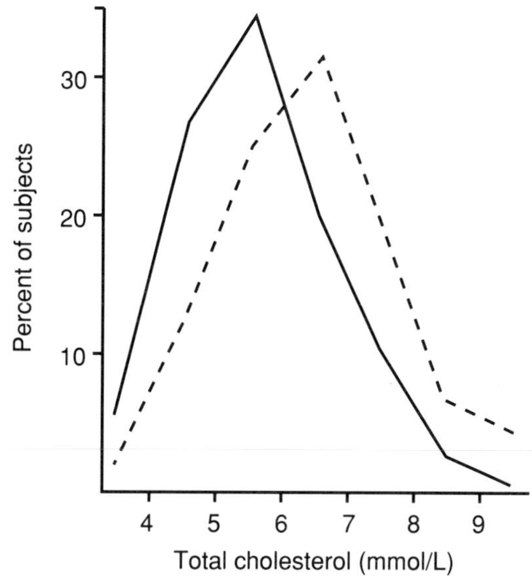

Fig. 1-2. Frequency distribution of serum total cholesterol in 1,923 men and women with DBP below 70 mmHg (solid line) and in 575 men and women with DBP above 99 mmHg (dashed line). The values have been age adjusted. (From Bønaa and Thelle,[10] with permission.)

IS THERE A TRUE BIOLOGIC ASSOCIATION BETWEEN CHOLESTEROL AND BLOOD PRESSURE?

Earlier studies of the possible association between serum cholesterol and BP have yielded different conclusions. This may be due to differences in the design of these studies and to differences in patient characteristics, such as age, sex, risk factor levels, and geographic areas. No correlation between BP and cholesterol was found in two studies of men over 50 years of age[11,12] and in a Danish study of both men and women no significant relationship was found at the age of 50 years.[13]

Stamler et al.,[14] using four different types of multivariate analyses, concluded that with the exception of white men aged 30 to 44 years, there were generally negative findings with respect to serum cholesterol and blood pressure. According to these authors this was in keeping with several other investigations.

The Oslo study investigated the cholesterol and BP relationship in more than 16,000 men aged 20 to 49 years.[15] A significant positive correlation was demonstrated with multivariate analysis. Body mass index and nonfasting serum triglycerides significantly influenced the relationship, whereas age, cigarette smoking, nonfasting blood sugar, season, socioeconomic status, and physical activity at work and leisure did not. The relationship between BP and cholesterol is further displayed in Figures 1-3 and 1-4, showing increasing serum cholesterol levels with increasing BP. The increase, however, is stronger in the younger age groups (corresponding to the Tromsö data), which may be one of the reasons why this association has attracted relatively little attention.

WHERE DO THESE ASSOCIATIONS LEAD?

If hypertension affected CHD risk mainly by modifying the more powerful influences of cholesterol on the atherosclerotic process, then, since the other etiologically more important risk factors would continue to act independently, a reduction of BP would not be expected to have a major impact on coronary risk. In terms of coronary prevention, it would be more important to lower total cholesterol than BP.

The true impact of the increased triglyceride level is difficult to assess,[16] but its significance is suggested by the strong negative association between HDL cholesterol and triglyceride as well as the atherogenic properties of lipoproteins such as very low-density lipoprotein (VLDL) remnants, intermediate-density lipoprotein, or abnormally small low-density lipoprotein

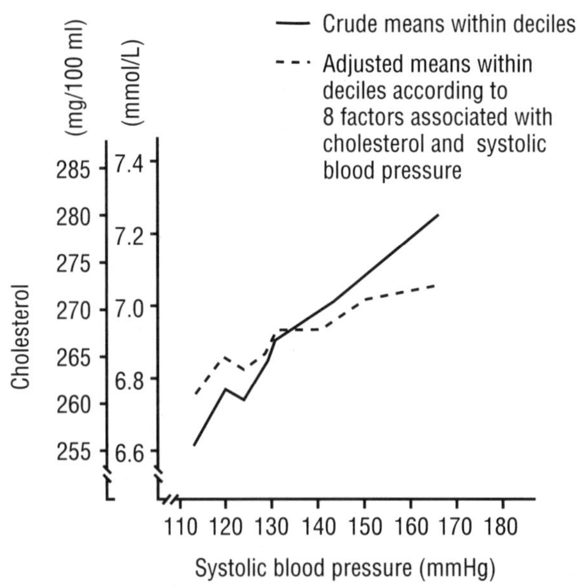

Fig. 1-3. Nonadjusted and adjusted correlation between systolic BP and serum cholesterol (adjusted for weight/height^2, serum triglycerides, glucose, age, cigarette smoking, seasonal variation, socioeconomic status, and physical activity at work or at leisure) in patients aged 40 to 49 years. (From Hjermann et al.,[15] with permission.)

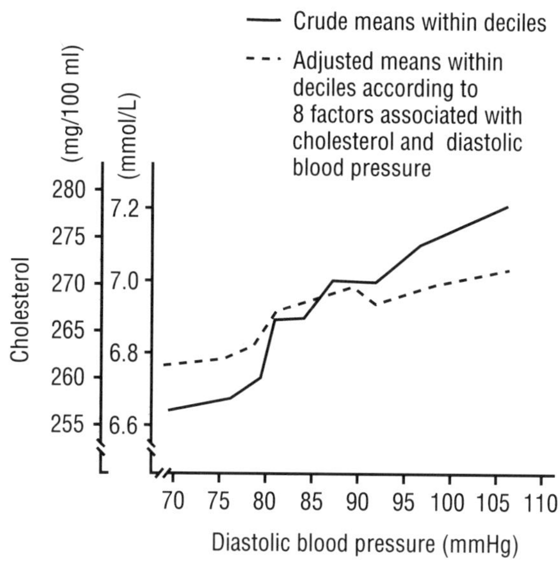

Fig. 1-4. Nonadjusted and adjusted correlation between DBP and serum cholesterol (adjusted for weight/height^2, serum triglycerides, glucose, age, cigarette smoking, seasonal variation, socioeconomic status, and physical activity at work or at leisure) in patients aged 40 to 49 years. (From Hjermann et al.,[15] with permission.)

particles; it may be that existing epidemiologic methods are too crude to disentangle the associations.[17]

Blood platelet dysfunction in hypertensive patients is another factor that has recently caught attention. Secretion of platelet-derived growth factor may contribute to the smooth muscle hyperplasia of atherosclerotic lesions. Thus, a number of mechanisms may be involved and antihypertensive drugs may not normalize these additional factors.

There is a possibility that a lack of CHD risk reduction could reflect an adverse effect of the drugs used. Both thiazides and β blockers have been reported to increase triglyceride levels and reduce the HDL cholesterol levels. The concomitant increase of blood glucose may also be a reflection of increased insulin resistance which has been reported to be related to increased secretion of VLDL.[18]

DO FISH OILS PLAY A ROLE IN HYPERTENSION AND HYPERTENSIVE COMPLICATIONS?

Our understanding of how blood pressure is related to CHD is sparse. It may be that hypertension needs the metabolic basis of a certain lipid level to become an atherogenic factor of importance, a view that was supported by a

Swedish study[19] where mild elevation of blood pressure carried little increase in CHD risk in the lowest quintile of the cholesterol distribution.

Another mechanism that has come more into focus in recent years is that a possible linkage between blood pressure and blood lipids could exist via the fatty acids.

From an epidemiologic point of view it has always been intriguing that populations with low coronary heart disease mortality have been reported to have high rates of stroke. This seemingly paradoxical situation has been observed in Japan, Hawaii, and California.[20]

This, in addition to the observation that the decline in stroke rates in industrialized countries started long before the decline in coronary mortality became evident, indicate that the two groups of disorders have different causes, even if high blood pressure is a risk factor for both. To explain these discrepancies it has been suggested that there are different causes of atherosclerosis in different cerebral arteries, e.g., that in the Japanese stroke is more often due to affections of the intracerebral arteries that again are related to blood pressure but not to the lipids.[21]

Obviously, the balance between all the factors related to atherosclerosis, i.e., clotting, thrombosis, and coagulation mechanisms, must be considered when between-population differences are to be explained. It is likely that populations with prolonged bleeding time are more subject to cerebral hemorrhages than other groups, who on the other hand will show relatively higher incidence of thromboembolic complications. In this respect the intake of fish and thereby of eicosapentaenoic and docosahexaenoic acids (n-3 fatty acids) may be of importance in explaining such between-population differences.

The relationship between fish and fish oils and blood pressure has been addressed in a number of studies, but differences in size, exposure to fish oils, as well as different result-variables make it difficult to draw firm conclusions.

The results of a recent randomized controlled trial involving 156 men and women given fish oil compared with corn oil showed a 4.6 mmHg decline in systolic and 3.0 mmHg decline in diastolic blood pressure.[22] The reductions were dependent upon the increase in platelet n-3 fatty acids, showing a strong dose-response relationship.

In addition, the triglyceride level was reduced in the n-3 users, consistent with what other studies have shown. The concomitant decline in triglycerides and blood pressure points toward the possible relationship between blood lipids and blood pressure.

The inter-relationship between blood pressure, blood lipids, and hemostatic factors as demonstrated by our fish oil trial is an argument for this hypothesis. It is probable that fish oils play a relatively minor role as a blood pressure determinant in populations with high salt intake, but they may influence hemostatic factors and thereby the incidence of hemorrhagic stroke in a hypertensive population.

The hypertensive patient represents a clinical challenge and until the mechanism of higher CHD risk are clarified, proper intervention is difficult to decide upon. The possible metabolic disturbances that are associated with high blood pressure per se and those that develop as a result of modern treatment must be taken into consideration when assessing the patient's overall risk for pressure-related complications.

REFERENCES

1. Report of Inter-Society Commission for Heart Disease Resources: Primary prevention of the atherosclerotic diseases. Circulation 42:A55, 1970
2. Study Group of the European Atherosclerosis Society: Strategies for the prevention of coronary heart disease: a policy statement of the European Atherosclerosis Society. Eur Heart J 8:77, 1987
3. Thelle DS: Hypercholesterolaemia. Importance as a coronary heart disease risk factor and rationale for treatment. Drug Investigation, suppl 2. 2:1, 1990
4. Holme I: Drug treatment of mild hypertension to reduce the risk of CHD: is it worth while? Stat Med 7:1109, 1988
5. Multiple Risk Factor Intervention Trial Research Group: Multiple risk factor intervention trial. Risk factor change and mortality results. JAMA 248:1465, 1982
6. Helgeland A: Treatment of mild hypertension: a five-year controlled drug trial. The Oslo study. Am J Med 69:725, 1980
7. Collins R, Peto R, MacMahon S et al: Blood pressure, stroke and coronary heart disease. Part 2. Short-term reductions in blood pressure; overview of randomized drug trials in their epidemiological context. Lancet 335:827, 1990
8. Løchen M-L: The Tromsø heart study: coronary risk factor levels in treated and untreated hypertensives. Acta Med Scand 224:515, 1988
9. Arnesen E, Thelle DS, Førde OH, Mjøs OD: Serum lipids and glucose concentrations in subjects using antihypertensive drugs: Finnmark 1977. J Epidemiol Community Health 37:141, 1983
10. Bønaa K, Thelle DS: Association between blood pressure and serum lipids in a population. The Tromsø study. Circulation 83:1305, 1991
11. Sive PH, Medalie JH, Kahn HA et al: Distribution and multiple regression analysis of blood pressure in 10,000 Israeli men. Am J Epidemiol 93:317, 1971
12. Castelli WP, Anderson K: A population at risk. Prevalence of high cholesterol levels in hypertensive patients in the Framingham study. Am J Med 80:23, 1986
13. Hagerup LM: Coronary heart disease risk factors in men and women. Acta Med Scand, suppl. 557:46, 1973
14. Stamler J, Rhomberg P, Schoenberger JA et al: Multivariate analysis of the relationship of seven variables to blood pressure: findings of the Chicago Heart Association detection project in industry, 1967-1972. J Chronic Dis 28:527, 1975
15. Hjermann I, Helgeland A, Holme I et al: The association between blood pressure and serum cholesterol in healthy men: the Oslo study. J Epidemiol Community Health 32:117, 1978
16. Austin MA: Plasma triglyceride as a risk factor for coronary heart disease. Am J Epidemiol 129:249, 1989

17. Grundy SM, Vega GL: Hypertriglyceridemia: causes and relation to coronary heart disease. Semin Thromb Hemost 14:149, 1988
18. Geaven GM: Looking at the world through LDL-cholesterol-colored glasses. J Nutr 116:1143, 1986
19. Samuelsson O, Wilhelmsen L, Andersson OK et al: Cardiovascular morbidity in relation to change in blood pressure and serum cholesterol levels in treated hypertension. Results from the primary prevention trial in Göteborg, Sweden. JAMA 258:1768, 1987
20. Gordon T: Mortality experience among the Japanese in the United States, Hawaii and Japan. Public Health Rep 72:543, 1957
21. Reed DM: The paradox of high risk of stroke in populations with low risk of coronary heart disease. Am J Epidemiol 131:579, 1990
22. Bønaa KH, Bjerve KS, Straume B et al: Effect of eicosapentanoic and docosahexanoic acid on blood pressure in hypertension. N Engl J Med 322:795, 1990

2

Glomerulosclerosis and Progressive Renal Disease

Peter C. Brazy
Jeffrey B. Kopp
Paul E. Klotman

INTRODUCTION

Glomerulosclerosis is the process whereby extracellular matrix materials replace functioning glomerular capillary tufts. Glomerulosclerosis is similar to atherosclerosis,[1] and the two processes coexist in many subjects. Both glomerulosclerosis and atherosclerosis of the renal arteries occur in the absence of clinical renal disease. The incidence of these processes is relatively frequent. For example, a recent prospective study reported that angiographic evidence of atherosclerosis of renal arteries was present in 30 percent of 1,303 patients who were having angiography for symptoms of coronary artery disease.[2] There were significant occlusions of the renal arteries in 15 percent and bilateral renal artery lesions in 4 percent of the cases.[2] The incidence of glomerulosclerosis has been estimated from renal tissue at autopsy. The number of sclerotic glomeruli appeared to increase with the age of the subject, and as many as 10 to 12 percent of all glomeruli were totally sclerotic in 70-year-old subjects.[3] The functional correlation of glomerulosclerosis and renal atherosclerosis is probably a decline in the glomerular filtration rate (GFR). There is a steady rate of decline in GFR with aging in normal men.[4] These observations suggest that pathologic alterations of the renal vascular tree are present but clinically undetected in many subjects. These lesions become clinically significant in patients who develop chronic renal insufficiency.

The two most common causes of end-stage renal disease (ESRD) in the United States are diabetes mellitus and hypertension. The major renal lesion in each of these diseases is glomerulosclerosis. A complete understanding of the pathogenesis of glomerulosclerosis in these disease processes may provide the critical step in developing strategies to treat progressive renal disease in patients. Data from animal and in vitro studies suggest that the pathogenesis of glomerulosclerosis may involve (1) physical factors such as increased glomerular capillary pressure,[5] (2) toxic factors such as serum lipids and free radicals,[6] and (3) regulatory hormones and growth factors that stimulate mesangial cell proliferation and production of extracellular matrix proteins.[7,8] Some of these factors may also affect the rate of atherosclerotic plaque formation in the systemic circulation.[1] Patients with diabetes mellitus and hypertension have markedly accelerated rates of atherosclerosis. The interrelationships between glomerulosclerosis, atherosclerosis, and progressive renal insufficiency are currently under investigation.

The purpose of this chapter is to review the natural history of progressive renal disease in patients, to identify factors that may affect the rate of decline in renal function (clinically), and to relate these factors to what is known about the pathogenesis of glomerulosclerosis. Our hypothesis is that factors regulating the process of glomerulosclerosis are important determinants of the rate of progressive renal insufficiency in patients.

CLINICAL ASPECTS OF PROGRESSIVE RENAL DISEASE

Renal Function in the Absence of Kidney Disease

Renal function in individuals without overt renal disease declines as a linear function of time. An age-related decline in GFR is evident from measurements of either inulin or creatinine clearance. A cross-sectional study of 548 healthy volunteers (the Baltimore Longitudinal Study of Aging) showed that creatinine clearance decreased in a linear manner at an approximate rate of 0.8 ml/min/1.73 m^2/yr[8] (Fig. 2-1). The subjects in this study were mostly white males. In this population there was a significant negative

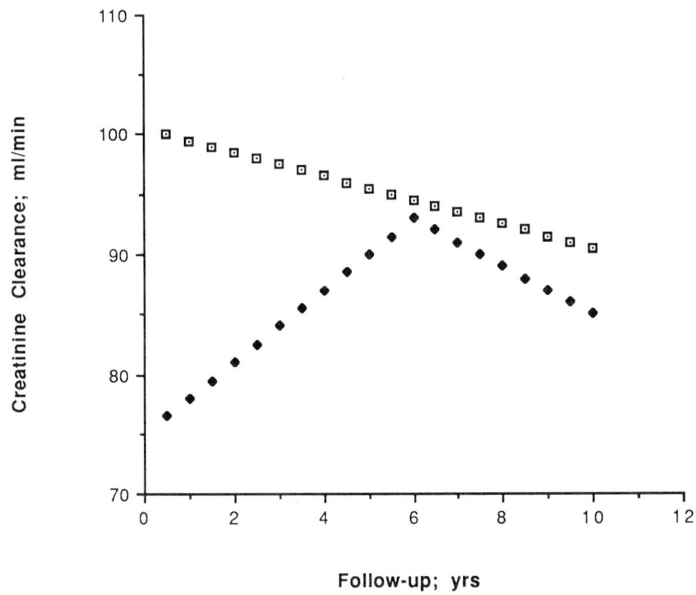

Fig. 2-1. Renal function in normal individuals and transplant recipients. Creatinine clearance is shown as a function of time for "normal" subjects (☐) and for patients who have received a successful renal transplant (◆). (Data from Lindeman et al.[9] and Brazy,[14] respectively.)

correlation between mean arterial blood pressure and the rate of decline in creatinine clearance.[9] Excluding subjects with possible renal pathology and individuals on antihypertensive medications, the presence of hypertension appeared to accelerate the rate of decline in creatinine clearance by 25 to 50 percent.[9] The decline of creatinine clearance with age in normal subjects is probably related to the development of glomerulosclerosis.[3] Thus, the presence of hypertension may accelerate the rate of glomerulosclerosis in individuals who, clinically, do not have renal disease.

Glomerular hyperfiltration may also affect the rate of decline in renal function. In animal models, hyperfiltration and increased glomerular capillary pressure have been associated with glomerulosclerosis and progressive renal insufficiency.[5] If the animal data are applicable to patients, than people with hyperfiltering kidneys should have more sclerotic glomeruli and a faster rate of decline in GFR. Observational data do not confirm this expectation. Individuals who donate kidneys develop hyperfiltration in their remaining kidney but do not appear to have an enhanced rate of decline in GFR. Several published studies showed that the value of serum creatinine in kidney donors did not change from the time of nephrectomy to 5,[10] 10,[11] or 15[12] years later. Serial measurements of GFR or serum creatinine in kidney donors indicate that GFR actually increases with time. Bonner and colleagues[10] measured GFR serially with iothalamate in 43 kidney donors over a 4-year period and found that GFR increased steadily. The rate of increase in GFR varied as a function of the age of the subject, but the absolute value of GFR increased in all cases. Serial measurements of kidney function (serum creatinine) in 66 kidney donors from the Mayo Clinic showed increasing renal function over a 5- to 10-year period.[13] The data indicate that hyperfiltration and hypertrophy of glomeruli occurred in these solitary kidneys. While there are no data on the incidence of glomerulosclerosis in these solitary kidneys, the stability of serum creatinine values in the reported studies suggests that the rate of glomerulosclerosis is not increased. The conditions present in the solitary kidney of transplant donors are associated with a lengthy period of increasing renal function and no evidence of an exaggerated reduction in renal function with age. In this setting, hyperfiltration is not the antecedent of progressive renal insufficiency.

The successful renal transplant recipient is another clinical example of reduced functional renal mass. We have recently examined the natural history of renal function in "successful" renal transplant recipients at the University of Wisconsin.[14] For the purposes of this study, a "successful" renal transplant was defined as a functioning kidney for more than 10 years with a stable serum creatinine. We identified 133 such patients from the records of the University of Wisconsin Transplant Service. Longitudinal data from these patients showed renal function (creatinine clearance) post-transplant gradually increased for an average of 6 years. After reaching a peak creatinine clearance value that averaged 90 ml/min for the group, the creatinine clearance declined at a rate of 2.0 ml/min/yr (Fig. 2-1). This rate of decline was about twice that observed in a normal population. The explanation for this accelerated rate of decline is unknown. These patients have increased levels of plasma cholesterol[15] and may have accelerated rates of atherosclerosis and glomerulosclerosis. All patients were taking prednisone, and increased glucocorticoid activity has been associated with a more rapid rate of progressive renal disease.[16] The presence of a mild form of chronic rejection or "transplant" glomerulonephritis cannot be excluded from the available data. Perhaps the same growth factors that stimulate functional hypertrophy for several years in the transplanted kidney produce a faster rate of glomerulosclerosis in the later years.

In the University of Wisconsin study,[14] the presence of diastolic hypertension or diabetes mellitus did not significantly affect the rate of decline in creatinine clearance. In another study[17] from the University of Minnesota, the glomerular pathology of biopsy specimens from kidneys transplanted into diabetic patients was correlated with several potential risk factors, such as blood pressure, glycemic control, and age of patient or donor. The only positive correlation was between the amount of mesangial expansion and glycemic control. These results in patients with transplanted kidneys differ from those observed in patients with chronic kidney disease. As described in the next section, hypertension and diabetes are factors that accelerate the rate of decline in renal function in patients with chronic renal insufficiency. Future investigations into the differences between transplanted kidneys and diseased, native kidneys may indicate which mechanisms are involved in progressive renal disease and glomerlosclerosis.

The Nature of Chronic Renal Disease

The natural history of renal function in patients with chronic renal disease appears to be one of progressive deterioration. Precise data on this point are limited. The rate of change in renal function should be determined from longitudinal measurements of GFR in a randomly selected sample of patients with renal insufficiency. The available data are from retrospective studies of selected patients with GFR estimated from serum values of creatinine. However, the conclusions derived from these observations are remarkably consistent. Most patients with renal insufficiency progress. In one example, 108 patients with mild to moderate renal insufficiency were followed for a mean duration of 41 months at a renal clinic in Liverpool, England.[18] Renal function was estimated from the reciprocal of serum creatinine[19] and 79 percent of these patients exhibited a deterioration of renal function.[18] In another example, 77 percent of 184 patients with renal insufficiency (serum creatinine greater than 1.5 mg/dl) who were followed for an average of 40 months in the Renal Clinic of the Durham VA Medical Center showed a significant rise in their serum creatinine (Brazy, unpublished observations). In a third example, 82 percent of 141 patients with renal insufficiency who were followed for an average of 47 months at the Nephrology Clinic of the University of Wisconsin–Madison showed a significant rise in their serum creatinine (Brazy, unpublished observations). Thus, retrospective data from three nephrology clinics that serve very different populations indicate that a high percentage of patients who present with renal insufficiency exhibit a progressive deterioration in renal function.

The true prevalence of progressive renal disease is unknown. The data cited above are based on selected populations. We do not know the natural history of renal function in patients with renal insufficiency who are not referred. In a survey of all patients who were treated for ESRD (dialysis or transplant) at the Durham VA Medical Center during a 6-year period (1982 through 1987), 56 percent of these patients had progressed to ESRD prior to their first visit to a nephrology clinic.[20] This observation suggests that a significant number of patients progress to ESRD outside the setting of a renal clinic.

The pattern of progressive loss of renal function may be episodic or continuous. Assume that renal function declines when the mechanisms of injury are active, and that renal function either improves or is stable when the mechanisms of injury are inactive. Then, disease processes that cause renal injury in an episodic manner will show decrements in renal function followed by partial recovery or periods of stable function. In this case, the time course of renal function would have a saw-tooth pattern (Fig. 2-2). Disease processes that continuously injure the kidney will have a time course of renal function

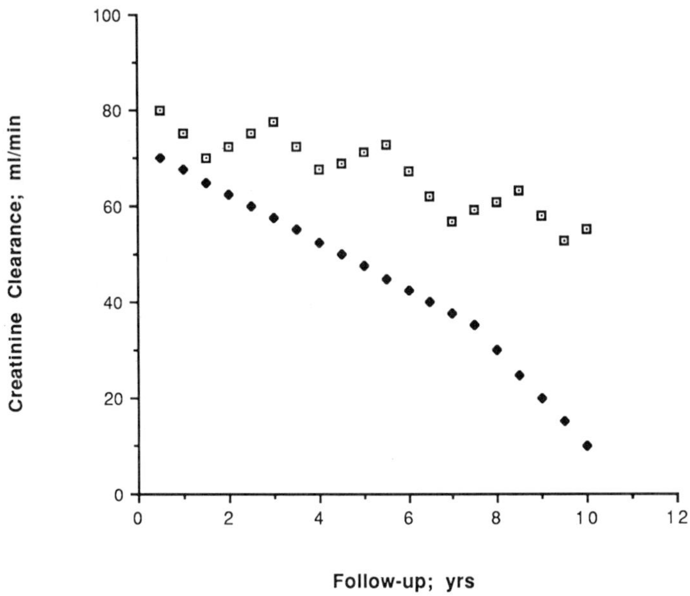

Fig. 2-2. Models of progressive renal insufficiency. These data are a hypothetic plot of creatinine clearances versus time for renal diseases that cause renal injury in either an episodic (⊡) or constant (◆) manner.

that has a downward slope which may be linear or curvilinear. With this very simplistic analysis we can examine plots of renal function versus time and determine whether the natural history of renal injury is episodic or continuous.

As mentioned above, there are no published studies in which precise measurements of GFR were made in a large number of patients with renal insufficiency over a long period of time. We do have data from studies in which measurements of serum creatinine were used to estimate renal creatinine clearances. Initial observations indicated that plots of reciprocal serum creatinine[19] or natural log values of serum creatinine[21] versus time were linear in some patients with renal insufficiency. A linear plot of reciprocal creatinine versus time implied a constant rate of decline in renal function, whereas a linear plot of the log of creatinine versus time implied loss of a constant fraction of remaining renal function.[22] The validity of such mathematical transformations of serum creatinine values as measurements of GFR in patients with renal insufficiency has been questioned.[23] The method may not provide a precise measurement of GFR in patients because advancing renal disease may induce changes in the rates of production and elimination of creatinine. However, this approach is readily available and inexpensive, and does provide some information about changes in renal function.[24]

When estimates of renal function have been used to examine the time course of progressive renal insufficiency, a downhill, linear pattern has frequently been observed. In the Renal Clinic in Liverpool, England, 85 of 108 patients had progressive renal insufficiency; 82 percent of these patients had a linear pattern to their rate of decline.[18] In the data from the Renal Clinic of the Durham VA Medical Center, the plot of reciprocal serum creatinine versus time was linear in 90 percent of the patients who were followed from early in their disease process stage until the time of dialysis.[20] On close inspection, the plots of renal function versus time were not always best described by a single line. In some patients, the time course could be described by a series of lines connected by break points. In almost every case, the lines had a negative slope, but at a break point, the negative slope became more or less steep. Analyses of changes in the slope of these plots indicate that interventions such as control of diastolic blood pressure,[20] modifications of dietary protein,[25] or simply more frequent visits to clinic[26] could slow the rate of decline in renal function. Together, these observations suggest that progressive renal insufficiency in patients is best characterized as a multifactorial process that is more or less continuously active.

The type of primary renal disease does not appear to be the main determinant of the rate of progressive renal insufficiency. Patients with the same diagnosis may have either a very slow or a quite rapid rate of progression. When examining a large number of patients, the average rates of decline in renal function in patients with primary hypertension, type II diabetes mellitus, chronic glomerulonephritis, or interstitial nephritis were not statistically different.[27,28] The degree of proteinuria may adversely affect the rate of progressive renal insufficiency. In the Liverpool Renal Clinic, patients with glomerulonephritis or diabetic nephropathy appeared to progress faster than patients with interstitial nephritis or polycystic kidney disease.[18] Younger patients (less than 45 years old) had faster rates of progressive renal disease than older patients.[28] This observation may relate to the degree of hypertension or the types of glomerulonephritis that are more frequent in this age group. Thus, factors characteristic of each patient are more important than the type of renal disease in determining the rate of progression.

The observational data suggest that the natural history of chronic renal disease is progressive deterioration. The processes causing the deterioration appear to be constantly at work. The effective rate of decline varies more from patient to patient than from one type of renal disease to another. The rate of decline may be modified by a single maneuver in patients with different types of renal disease. Therefore, some of the mechanisms that are responsible for the loss of renal function may be common to the condition of renal insufficiency and independent of the primary cause of renal injury. The processes of vascular injury and glomerulosclerosis are potential explanations for these observations.

Essential Hypertension and Glomerulosclerosis

Hypertension is associated with an accelerated rate of decline in renal function in "normal" subjects[9] and in patients with pre-existing renal disease.[20] The pathologic effects of hypertension on renal blood vessels include hypertrophy of the myointima in small intrarenal arteries (interlobular arteries) and hyaline sclerosis of afferent arterioles. These processes lead to a reduction in blood flow to glomeruli and tubular segments. With chronic hypertension, glomerular capillaries show characteristic wrinkling and thickening of basement membranes. There is an associated increase in mesangial matrix and collagen within Bowman's capsule. Eventually, the glomerular structures are completely sclerotic. This is followed by ischemic damage to renal tubules and interstitial fibrosis.

Hypertension is frequently associated with the processes that cause renal insufficiency. Data from a large hypertension clinic in Indianapolis indicate that 18 percent of the 6,880 patients developed renal insufficiency (a plasma creatinine of greater than 2 mg/dl) during the follow-up period, which averaged 5.2 years.[29] Black patients had twice the risk of developing renal insufficiency as white patients. Similar data were reported from Birmingham, Alabama. Rostand and colleagues[30] followed 94 patients who initially had a normal serum creatinine and treated hypertension. During follow-up, which averaged 58 months, 15 percent of the patients showed an increase in serum creatinine despite "good" blood pressure control. Black patients were twice as likely as white patients to have progression of renal insufficiency.[30] These data suggest that renal insufficiency develops frequently in the setting of hypertension. This development may occur even when blood pressure has been "adequately" controlled.

The predilection of black patients to develop renal insufficiency suggests the presence of genetic factors that influence the rate of glomerulosclerosis. The risk of developing ESRD was much greater in blacks than whites.[31] This increased risk exceeds that expected from the higher incidence of hypertension in blacks.[32] The studies cited above[29,30] indicated that black patients with hypertension were more likely than white patients to develop renal insufficiency. Once renal insufficiency is present, however, the rate of change in renal function was not affected by race. Brazy and Fitzwilliam[28] studied the records of 200 patients with chronic renal insufficiency. They reported that the rate of progressive renal disease was not different between black and white patients when comparing races in both normotensive and hypertensive groups. Thus, black patients with hypertension have a predilection to develop renal insufficiency and glomerulosclerosis. However, this predilection does not cause the process of renal insufficiency to progress more rapidly to ESRD.

The effect of hypertension on changes in renal function was documented during early studies with antihypertensive treatment. In the 1950s, Moyer and colleagues[33] performed inulin clearances in 133 patients with hypertensive vascular disease. Serial measurements of renal function were performed in 64 patients (45 treated and 19 untreated). The GFR declined by an average of 13 ml/min/yr in patients with untreated, severe hypertension ($N = 11$). With treatment, this rate of decline was reduced to 1.4 ml/min/yr ($N = 31$). In patients with untreated, moderate hypertension, GFR declined by

5 ml/min/yr ($N = 8$); in patients with treated, moderate hypertension, GFR declined by 3 ml/min/yr ($N = 14$). Thus, the rate of progressive renal disease or glomerulosclerosis appears to depend on the presence and severity of hypertension.

A question that is being addressed by clinical studies is whether the adverse effect of systemic hypertension on progressive renal disease is due solely to the elevation of blood pressure or to other factors that are involved in the pathogenesis of hypertension. The data described above would support the first alternative because hypertensive black and white patients have similar rates of progressive renal disease despite different mechanisms of hypertension. Another approach would be to determine if specific antihypertensive medications slow the rate of progressive renal disease more effectively than other medications at equivalent levels of blood pressure. In studies in which various types of antihypertensive medication were compared, angiotensin-converting enzyme inhibitors,[34,35] calcium channel blockers,[28,36] and minoxidil[28,37] appeared to preserve kidney function better than other antihypertensive medications. These observations need to be confirmed by well-controlled, prospective clinical trials. However, the implication of such observations is that specific therapy can reduce the rate of decline in renal function by blocking a pathway involved in the process of glomerulosclerosis.

Diabetes Mellitus and Glomerulosclerosis

Diabetes mellitus affects smaller blood vessels, such as the glomerular capillary bed. The early glomerular lesions of diabetic nephropathy are expansion of the mesangial matrix and increased thickness of the glomerular basement membrane. These changes are followed by restriction of glomerular capillary lumen and a reduction in the surface area of the glomerular filter. Adaptations to this situation include hypertrophy of remaining renal glomeruli and increased intraglomerular pressure across remaining filtration surfaces. These changes are associated with alterations in glomerular permeability to large molecules. Hypertension, alterations in plasma lipids, and release of locally active vasoregulatory substances may also contribute to the ongoing process of injury. The end result is sclerotic glomeruli and a decline in functional renal mass.

While diabetic nephropathy is one of the leading causes of ESRD in the United States, only a fraction of patients with diabetes mellitus develop clinically significant renal disease. Humphrey and colleagues[38] from the Mayo Clinic determined the incidence of renal failure in patients with either non-insulin-dependent diabetes mellitus (NIDDM) or insulin-dependent diabetes mellitus (IDDM) from a retrospective cohort study. Of 1,832 patients with NIDDM who initially did not have renal failure, 25 developed renal failure. This represents an incidence of 133 per 100,000 person-years, which was comparable to that observed in IDDM patients (177 per 100,000 person-years). In the NIDDM group, the risk of renal failure was associated with the presence of persistent proteinuria at the time of diagnosis of NIDDM. In a European study,[39] the cumulative risk of developing proteinuria after a 20-year history of diabetes was approximately 28 percent for both IDDM and NIDDM patients. The cumulative risk of developing an elevated serum creatinine after 5 years of proteinuria was approximately 60 percent for both types.

The appearance of familial clustering in diabetic nephropathy suggests that genetic factors may control its development. Seaquist and colleagues[40] studied concordance rates for diabetic nephropathy in two sets of families in which probands and siblings had diabetes mellitus. In probands without diabetic nephropathy, only two of 12 diabetic siblings had nephropathy. In probands with nephropathy, 24 of 29 diabetic siblings had nephropathy. In another study, the sodium–lithium countertransport mechanism in the erythrocyte has been suggested as a genetic marker for diabetic nephropathy and hypertension.[41] In 89 patients with type I diabetes mellitus, the indicators for development of nephropathy (33 patients) were a hypertensive parent, a high maximal rate of red blood cell sodium–lithium countertransport, and poor glycemic control. Recently, Gall et al[42] reported that patients with diabetes mellitus (either type I or II) had high rates of red blood cell sodium–lithium countertransport compared with normal individuals. However, they found no differences in the activity of this transport mechanism between patients with and without nephropathy. Thus, the epidemiologic data suggest a familial pattern in diabetic nephropathy, but a genetic mechanism has not been identified. An atherogenic lipoprotein profile has also been reported in diabetic patients with increased red blood cell sodium–lithium countertransport.[43]

Hypertension in patients with diabetic nephropathy appears to contribute greatly to the rate of decline in renal function. Early studies by Mogensen[44] and Parving et al.[45] indicated that correction of hypertension was associated with a marked slowing of the rate of decline in GFR. In Parving et al's study of 10 patients with type I diabetes mellitus, the pretreatment rate of decline in GFR was 10.8 ml/min/yr. After control of blood pressure with metoprolol, hydralazine, and furosemide, urinary albumin excretion was decreased and the rate of decline in GFR was reduced to 3.6 ml/min/yr.[45] This study is remarkable for the 67 percent change in the rate of decline in GFR. Was the effect due to the reduction in blood pressure alone, or to the choice of antihypertensive agents? The study by Brazy and Fitzwilliam[28] did not find any special benefit of beta blockers or hydralazine in reducing the rate of decline in renal function, but their patient population was very different. Thus, the question remains unanswered. Other antihypertensive medications, such as the angiotensin-converting enzyme inhibitors, are being examined to determine if they specifically block the mechanisms of glomerular injury in patients with diabetes mellitus. At this time, it appears that control of hypertension is the most important factor in patients with diabetic nephropathy.

MOLECULAR BASIS OF GLOMERULOSCLEROSIS

A therapeutic goal in patients with renal insufficiency is to block the rate of glomerulosclerosis. Information is rapidly accumulating on the character and function of the extracellular matrix materials that are the building blocks of glomerulosclerosis. Application of this information may provide therapeutic options for the treatment of progressive renal insufficiency. The following section is an overview of recent advances in our knowledge of renal extracellular matrix proteins.

Components of Renal Extracellular Matrix

From the clinical studies discussed above, it is apparent that the hallmark of progressive glomerular disease is a fall in GFR associated with deposition of extracellular matrix proteins in glomerular capillary loops and mesangium. Yet, the exact mechanism responsible for the accumulation of matrix proteins in the glomerulus remains unknown. It is likely that a common pathogenetic pathway exists that either increases transcription of the genes encoding the extracellular matrix proteins, enhances translation of the proteins, or inhibits their degradation. Hemodynamic alterations have been proposed as a common pathway responsible for the relentless progression of renal insufficiency to ESRD,[5] but a direct etiologic relationship between changes in capillary plasma flow or transcapillary pressure and extracellular matrix deposition in the glomerulus remains unproven. In fact, several lines of evidence suggest that hemodynamic changes are not causally related to experimental glomerulosclerosis. There is dissociation between early hemodynamic changes and subsequent glomerulosclerosis in the remnant nephron model, and angiotensin-converting enzyme inhibition prevents glomerulosclerosis without altering transglomerular capillary pressure.[46,47] Similarly, in uninephrectomized rats, high protein intake is associated with accelerated glomerulosclerosis without altering transglomerular capillary pressure or further increasing single nephron plasma flow.[48] As an alternative to hemodynamic changes, recent evidence points to metabolic factors, such as the generation of locally acting cytokines, growth factors, and inflammatory mediators, as the primary determinants that shift resident glomerular cells to a sclerosing phenotype. To better understand the abnormal expression of the extracellular matrix proteins in progressive renal disease, it is important to understand their normal control during renal development.

The extracellular matrix of the undifferentiated interstitial mesenchyme of the developing mammalian metanephros contains fibronectin and the interstitial collagens types I and III.[49] The invasion of the branching ureteric bud into the mesenchyme induces the mesenchymal cells to undergo epithelial differentiation and to express basement membrane components, as shown in vivo[49] and in organ culture.[50,51] The glomerular basement membrane arises as a product of both the endothelium and the visceral epithelium, resulting in a fused double basement membrane. With continued nephron growth and development, the epithelial podocytes synthesize additional basement membrane components that are spliced into the existing basement membrane.[52] Synthesis of basement membrane components declines markedly after birth in the mouse, and presumably in the human, but a low level of synthesis persists in older animals and may represent constitutive remodelling of the glomerular and tubular basement membranes.[53]

The molecular components of basement membranes have been reviewed recently[54,55] and are listed in Table 2-1, which shows the major extracellular matrix molecules comprising the glomerular basement membrane and mesangium. The apparent molecular weight of the intact molecules and their subunits are derived from migration on Laemmli sodium dodecyl sulfate polyacrylamide gel electrophoresis (SDS-PAGE) gels. Typical values for the molecular mass (Mr) have been selected, with the recognition that the reported sizes differ between studies depending on animal species, experimental system (in vivo versus in vitro), and the particular features of the SDS-PAGE system used. Furthermore, collagens are nonglobular proteins and, therefore, the values for Mr obtained by SDS-PAGE are particularly discrepant from the molecular mass values obtained by density sedimentation. The subunit Mr refers, in the case of glycoproteins, to subunit size after reduction and Laemmli SDS-PAGE, and in the case of proteoglycans, to core protein size after digestion with glycosaminoglycan-degrading enzymes.

Table 2-1. Major Extracellular Matrix Components of the Glomerulus

Component	Mr ($\times 10^{-3}$ kd)	Subunits	Subunit Mr ($\times 10^{-3}$ kd)	GAG chains	Chromosome (Human)
Integral to basement membrane					
Type IV collagen	550–600	α1(IV)	200		13q33-34
		α2(IV)	200		13q33-34
		α3(IV)	200		ND
		α4(IV)	200		ND
		α5(IV)	200		Xq22
Laminin	800	A	400		18p11.3
		B1	200		7q22
		B2	200		1q25-31
		Merosin	ND		ND
		S-laminin	190		ND
Entactin/nidogen	150				ND
Heparan sulfate PG	400		Core 250	3–4	ND
Other normal glomerular components					
Fibronectin	560		230–270		2p323-qtr
Thrombospondin	450		185		15q15
Chondroitin sulfate PG	500		150		ND
Biglycan (PG-I)	230		Core 45	2	Xq13-qter
Limited to the sclerotic glomerulus					
Type III collagen		α1(III)			2q24.3-31
Decorin (PG-II)	150		Core 45	1	12p12.1-qter

Abbreviations: Mr, apparent molecular weight; GAG, glycosaminoglycans; PG, proteoglycan; ND, not determined.

Type IV collagen is quantitatively the major component of basement membranes and provides underlying structure. Each type IV collagen trimer interacts with its neighbors via globular N-terminal and NC1 domains to form stable mesh-like networks. The classical type IV collagen heterotrimer is composed of two α1(IV) and one α2(IV) chains.[56] Monoclonal antibodies directed against the collagen α1(IV) and α2(IV) chains localize to glomerular and tubular basement membranes, Bowman's capsule, and mesangium.[57]

It has been known for some time that collagenase digestion of type IV collagen yields noncollagenous peptides that are not contained within the α1 or α2 chains. Recently, three novel type IV collagen chains have been described, designated α3, α4, and α5. The nature of the collagen molecule(s) in which these chains appear remains to be determined. The NC1 domain of collagen α3(IV) reacts strongly with serum containing Goodpasture's antibody, suggesting that it is the Goodpasture antigen.[58] The bovine collagen α3(IV) gene has recently been cloned and shown to have considerable homology to the α1 and α2 chains,[59] while only partial amino acid sequence is known for collagen α4(IV).[60] The gene-encoding collagen α5(IV) is located on the X chromosome and has considerable homology with the α1 chain (83 percent) and the α2 chain (63 percent).[61] Mutations in the α5 gene have been incriminated in Alport's syndrome.[62,63] Interestingly, there is a distinct distribution of type IV collagen containing the α1 and α2 chains, the α3 and α4 chains, and the α5 chains in the developing human kidney. Polyclonal antibodies directed against type IV collagen react with all basement membranes, including those of the ureteric bud, S-shaped tubule, glomerulus, and capillaries. The α3 and α4 chains appear in the glomerulus only after capillary invasion has occurred, while α5 is expressed in the ureteric bud and the glomerulus after capillary invasion, but not in more primitive glomeruli.[64] Antibody raised against a synthetic peptide from collagen α5(IV) stains glomerular basement membrane but not Bowman's capsule or tubular basement membrane in the adult kidney.[61] Although preliminary, these studies suggest that type IV collagen molecules may exist in several isoforms with distinct distribution and raise the possibility that these isoforms have distinct functions within the kidney. Type V collagen has been variously reported to be present in the glomerular basement membrane or the renal interstititum.[65,66] In addition, as yet uncharacterized collagenous molecules may also be present in the glomerular basement membrane.[65,67]

Laminin is a large glycoprotein (apparent Mr, 800 kd), which is a major component of all basement membranes. Among the basement membrane proteins, laminin has the most diverse biologic functions, with epitopes that bind to other matrix proteins and that are responsible for cell attachment.[68] Laminin was originally isolated from the Englebreth-Swarm-Holm (EHS) rat sarcoma, where it is a heterotrimer composed of a large A chain and two smaller B1 and B2 chains. All three chains have considerable homology, and the homology between the B1 and B2 chains is particularly close. The three chains are linked by disulfide bonds to form a cruciform structure as visualized in electron microscopic studies.

The genes for the laminin A, B1, and B2 chains are coordinately expressed in EHS sarcoma but not in either newborn or adult kidney. Thus, murine and rat kidney express little A chain mRNA and substantially more B1 than B2 chain mRNA, as shown by Northern analysis.[69,70] The expression of B1 and B2 chains and the paucity of A chain in the kidney has been confirmed by in situ hybridization.[53] Using immunolocalization, Sanes and colleagues[71] reported that laminin B1 is found in tubular basement membrane and to a lesser extent in mesangium, while it is absent from glomerular basement membrane. Laminin B2 is expressed in both glomerular and tubular basement membranes; laminin A has a similar distribution but the staining is less intense.[71] On the other hand, laminin A chain mRNA is strongly expressed in the comma- and S-shaped tubules of the embryonic mouse kidney.[51] These observations suggest that a variety of laminin molecular isoforms exist, with different tissue distribution and, likely, different functions.

Recently, other matrix molecules have been described that share homology with laminin and appear to represent additional laminin chains. These include merosin, an analog of the A chain, and S-laminin, an analog of the B1 chain.[71] While merosin is not expressed in kidney, antibodies directed against S-laminin react with glomerular basement membrane more intensely than tubular basement membrane.[71] Thus, it appears that different laminin isoforms, composed of different chains, may occur at different locations in the kidney. As a basis for future investigation, a hypothesis concerning the nature of these laminin molecules in the kidney might be as follows: A-S-B2 in the glomerular basement membrane and A-B1-B2 in the tubular basement membrane. Since laminin in the mesangium appears to contain chiefly B1 chain, it is unclear what its structure might be. Alternate possibilities include dimers composed of the two smaller chains and the existence of as yet uncharacterized laminin chains. Expression and assembly of laminin chains likely differ in embryonic, adult, and diseased kidneys, and will doubtless be a fruitful area of investigation.

Entactin/nidogen and heparan sulfate proteoglycan are also integral constituents of renal basement membranes. Entactin/nidogen is a 150-kd protein that forms calcium-dependent complexes with laminin. These complexes have a 1 : 1 stoichiometry in EHS tumor.[72] Immunolocalization studies suggest that entactin/nidogen is present in both glomerular capillary loops and mesangium.[73] Different tissues express extracellular matrix heparan sulfate proteoglycans with core proteins of various sizes but which nonetheless are immunologically related and may arise by proteolytic processing of a larger precursor. The EHS heparan sulfate proteoglycan has a 400-kd core protein and two to three heparan sulfate glycosaminoglycan side chains.[74] In the rat glomerular basement membrane, the heparan sulfate proteoglycan has an assortment of core protein sizes, ranging from 150 to 400 kd, with the 250-kd size being most prevalent. All of these core proteins are recognized by antibody directed against the core protein of the EHS heparan sulfate proteoglycan.[75] A portion of the sequence of the gene encoding the mouse heparan sulfate proteoglycan has been reported.[76]

The glomerular mesangium is a specialized extracellular matrix compartment that surrounds the mesangial cell and extends to the glomerular basement membrane and the endothelial cell membrane. The normal constituent molecules of the mesangial matrix include type IV collagen, laminin, and entactin/nidogen. Among the basement membrane components, only heparan sulfate proteoglycan is absent from the mesangium, as suggested by immunohistochemical technique.[77] Fibronectin is present in the glomerulus, but it is unclear whether it is a component of the glomerular basement membrane or is limited to the mesangium.[78] In addition, the mesangium contains thrombospondin[79] and several chondroitin sulfate proteoglycans. An antibody to a basement membrane chondroitin sulfate proteoglycan (intact size, 500 to 600 kd; core protein, 150 to 160 kd) localizes to tubular basement membrane and mesangium but not glomerular basement membrane.[80] Biglycan is a small proteoglycan that is expressed in many tissues and can bear either chondroitin sulfate or dermatan sulfate glycosaminoglycan side chains. Immunolocalization studies in the normal mouse indicate that biglycan is expressed by the glomerular endothelium and to a more limited extent in the mesangium.[81]

Renal Matrix Alterations in Glomerulosclerosis

Progressive renal disease of many etiologies is characterized by increased accumulation of acellular material within the glomerular mesangium. Initially, this process is characterized by focal areas of glomerular hyalinosis, acellular material that stains with eosin but not with periodic-acid Schiff, together with capillary collapse and adhesions of the glomerular tuft to Bowman's capsule. With time, the mesangial compartment is occupied by material that stains positively with both periodic acid-Schiff and silver. Ultimately, the capillary tuft is replaced by this sclerotic material and ceases filtration.[82] Other features of progressive glomerulosclerosis are more variable. The mesangial compartment is in some cases occupied by an increased number of cells, both mesangial cells and resident macrophages. In the case of diabetic nephropathy, there is an associated thickening and wrinkling of the glomerular basement membrane, which contributes to compromised capillary lumens. In addition to an expanded mesangial compartment, some patients with diabetic nephropathy also exhibit a characteristic nodular glomerulosclerosis (Kimmelstiel-Wilson lesion).

The mesangial components that are expressed as a consequence of inflammation or the sclerotic process belong to two classes. First, normal mesangial matrix molecules accumulate in excess quantities. These include laminin, type IV collagen, heparan sulfate proteoglycan, and fibronectin. Second, the sclerotic mesangium contains matrix molecules that are not usually present in this location. These include interstitial collagens, particularly type III collagen, and the small proteoglycan decorin. These matrix constituents are typical products of fibroblasts and related cell types. In a sense, the reappearance of these interstitial matrix products represents a return to biosynthetic activity of the embryonic renal interstitial mesenchyme. Cultured mesangial cells produce matrix proteins characteristic of both basement membrane and the interstitium,[83] suggesting that the mesangial cell in the tissue culture environment may have an activated, remodelling/sclerosing phenotype.

In patients with focal glomerulosclerosis, immunolocalization studies indicate that type IV collagen appears predominantly in the expanded mesangium, while type III collagen appears in glomerular synechiae, areas of periglomerular fibrosis, and the fibrotic interstitium. The distribution of type IV collagen in the uninvolved mesangium and glomerular capillary loops is normal.[84] Studies of matrix protein distribution in various primary renal diseases suggest that the proteins expressed and the regional pattern of expression differ considerably between different renal diseases.[85,86] Nonetheless, with progressive glomerulosclerosis, particularly with a mesangial proliferative component, the mesangium contains increased amounts of fibronectin, laminin, heparan sulfate proteoglycan, and types III and IV collagen.

Despite the widespread use in experimental animals of the nephron ablation model of progressive glomerulosclerosis, relatively little is known about the nature of or the mechanisms responsible for the accumulated glomerular matrix components. In rats subjected to experimental membranous nephropathy and unilateral nephrectomy, the glomerulosclerosis is characterized by the accumulation of type IV but not type III collagen within the mesangium.[87] The use of transgenic animals as models of renal disease offers the potential to investigate the effect of particular genes on the process of glomerulosclerosis. For example, mice transgenic for the large T antigen of the SV40 virus develop glomerulosclerosis, although only a few glomeruli progress to obsolescence.[88] Similarly, mice transgenic for growth hormone develop mesangial proliferation, glomerular synechiae, and focal (and occasionally global) glomerulosclerosis.[89,90] Mice transgenic for the envelope and regulatory proteins of the human immunodeficiency virus type 1 develop ESRD and glomerulosclerosis that is initially focal and segmental in distribution but that progresses to more diffuse and global involvement. Immunofluorescence studies of the expanded mesangium of affected glomeruli demonstrate increased quantities of type IV collagen, laminin, heparan sulfate proteoglycan, and fibronectin within the mesangium.[91]

Alterations in the renal expression of chondroitin sulfate proteoglycans are another component of the sclerosing process. As indicated above, the normal glomerulus expresses the small proteoglycan biglycan. In experimental glomerulonephritis induced by antibodies to the Thy-1 antigen, there is increased synthesis of biglycan as well as of another related proteoglycan, decorin, by isolated glomeruli.[92] Activated mesangial cells are a possible source for these proteoglycans. Cultured human mesangial cells release two distinct chondroitin/dermatan sulfate proteoglycans that have core proteins of approximately 45 kd and that are identified by immunologic techniques as biglycan and decorin.[7,93]

Diabetes mellitus in humans and experimental animals is associated with evidence of dysregulation of renal extracellular matrix protein expression. More is known about the matrix alterations in this disorder than other forms of glomerulosclerosis, perhaps in part due to the availability of animal models. Biochemical investigation has shown increased accumulation of basement membrane components in diabetic glomeruli from experimental animals. Thus, glomeruli from streptozotocin-diabetic rats contain more type IV collagen compared with glomeruli from normal rats.[94] On the other hand, the data for glomerular content of heparan sulfate proteoglycan are conflicting, with an increase reported in streptozotocin-diabetic rats[95] and a decrease reported in diabetic human patients.[96] This discrepancy underscores the need for caution in extrapolating from studies of diabetic rodents to mechanisms of human disease.

Immunohistochemical investigation of the sclerotic regions of human diabetic glomeruli show evidence of an increase in laminin, fibronectin, and types IV and V collagen.[97–99] Type IV collagen is increased in the glomeruli obtained from streptozotocin-diabetic rats, and insulin treatment is associated with both normalization of type IV collagen expression and the appearance of type III collagen in the mesangium.[100] The glomerular mesangium of obese, insulin-treated diabetic rhesus monkeys expresses increased amounts of type IV collagen, laminin, and fibronectin[101] (Kopp, unpublished data). Differential expression of type IV collagen chains has recently been reported in human patients with insulin-dependent diabetes and nephropathy. The thickened glomerular basement membrane reacts strongly with antibodies directed against α3(IV) and α4(IV), and reacts to a much lesser degree with antibodies against classic type IV collagen, including α1(IV) and α2(IV). By contrast, the expanded mesangium contains mainly classic type IV collagen. Obsolescent glomeruli contain exclusively α3(IV) and α4(IV) chains.[102] These intriguing findings suggest differential regulation of type IV collagen isoforms in disease, as was discussed earlier in the developing nephron.

Recently, several studies of diabetic rodents have shown increased steady-state levels of mRNA for extracellular matrix proteins, although there are important differences depending on the model used. The steady-state levels of laminin B1, collagen α1(IV), and fibronectin mRNA are increased in the kidneys obtained from streptozotocin-diabetic rats.[103,104] In the genetic diabetic KKay mouse model of NIDDM, steady-state levels of laminin B1 chain and heparan sulfate proteoglycan mRNA are unchanged while collagen α1(IV) mRNA is increased.[105] The reasons for these discrepant findings concerning the changes in laminin B1 chain mRNA are not immediately apparent, but may be due to differences between streptozotocin-induced diabetes and the genetic obese models of diabetes. Whether the observed increases in the amounts of steady-state mRNA are due to enhanced transcription or enhanced mRNA stability remains to be determined. If an increased gene transcription is found, then an important research objective will be the identification of nuclear transcription factors that regulate matrix gene expression in the diabetic kidney.

Proper temporal and spatial regulation of gene expression requires that the initiation of gene transcription be precisely controlled. This control is exercised in part by certain DNA-binding proteins, or transcription factors. Binding sites for transcription factors exist in the DNA both within the promoter, immediately 5′ to the transcription start site, and in enhancer and silencer elements. The promoter elements must be positioned at a specific site adjacent to the transcription start site and function only in one orientation, while the enhancer and silencer elements are designated as those that act independently of position, distance, and orientation. The regulatory regions and the transcriptional factors that bind to these regions remain to be defined for most basement membrane proteins.[106] In particular, little is known about transcription factors that might be specific to the kidney. In preliminary work from our laboratory, electrophoretic mobility shift assays using fragments of the mouse collagen α1(IV) promoter and enhancer and nuclear extracts from human and mouse renal tissue indicate specific binding by as yet unidentified DNA-binding factors. These complexes differ in molecular mass from those in EHS tumor extract, suggesting the possibility of tissue-specific transcriptional factors.[107]

Mechanisms of Altered Glomerular Matrix Gene Expression

During nephronogenesis, subsets of basement membrane genes are expressed in a coordinated fashion in different portions of the nephron.[51] After nephronogenesis is complete and during adult life, matrix proteins are synthesized at low levels, presumably representing ongoing replacement of basement membrane and mesangial matrix. A common feature of response to injury or inflammation by many tissues is the increased deposition of matrix proteins. In general, this tissue response has both tissue-specific and generalized features. The kidney is exquisitely sensitive to malfunction as a consequence of scar formation stemming from the repair process. Whereas many mesenchymal tissues (skin, bone, muscle, liver) and certain epithelial tissues (spleen, intestine) can maintain function despite a certain degree of structural disorganization, normal renal function is predicated on the precise tissue structure. Thus, scarring anywhere along the nephron, from glomerulus to renal pelvis, threatens the filtration process. Further, the kidney has the added disadvantage that hypertrophy of remaining nephrons to compensate for tissue damage may further damage those nephrons. For these reasons, the normal damage repair process that is adaptive in many tissues may be maladaptive in the kidney and may contribute to the pathogenesis of ESRD.

A wide variety of potential mediators of the renal inflammatory response has been described.[7,8,108] A complete listing would be beyond the scope of this chapter, but these may include peptides (interleukins 1 and 6, tumor necrosis factor, transforming growth factor-b, granulocyte-macrophage colony-stimulating factor, epidermal growth factor, and platelet-derived growth factor) and lipids (eicosanoids and platelet-activating factor). Whereas a number of these factors have been shown to participate in the glomerular cell proliferation and matrix synthesis in experimental glomerulonephritis, less is known about the role of these factors in progressive glomerulosclerosis.

Several recent studies have investigated factors that might regulate renal matrix production in glomerulosclerosis. Glomeruli from diabetic animals have an altered eicosanoid profile, including increased release of PGE_2, 6-keto$PGF\alpha$, and thromboxane.[109] There is evidence suggesting that the vasoconstricting eicosanoid thromboxane may serve to increase basement membrane gene expression in renal cells. Thromboxane increases type IV collagen and laminin synthesis by cultured mouse teratocarcinoma F9 cells and human fetal mesangial cells.[107] Inhibiting the renal synthesis of thromboxane A_2 with the specific thromboxane synthetase inhibitor U-63557A prevents the increase in collagen α1(IV) mRNA that is observed in the obese diabetic KKay mouse.[105] Similarly, inhibition of thromboxane synthesis ameliorates the functional and pathologic changes of progressive glomerulosclerosis in the remnant kidney model.[110,111] Administration of both anticoagulant and non-anticoagulant heparins retard the progression of glomerulosclerosis; these proteoglycans may act by binding to and inhibiting the action of growth factors on renal cells.[112]

Systemic metabolic abnormalities in diabetes may also affect extracellular matrix gene expression. EHS tumor grown subcutaneously in genetically diabetic db/db mice contains more laminin and less heparan sulfate proteoglycan compared with tumor grown in nondiabetic mice, suggesting that systemic factors may play a role in the accumulation of basement membrane proteins.[113] Furthermore, glucose regulates the synthesis of matrix proteins by cultured cells. Thus, elevated medium glucose increases mRNA expression of type IV collagen and fibronectin but not type I collagen in cultured human umbilical vein endothelial cells[114] and increases accumulation of fibronectin, laminin, and type IV collagen and fibronectin mRNA in cultured rat mesangial cells.[115]

In summary, while our understanding of the protein constituents of the sclerotic process has improved, the molecular mechanisms responsible for their accumulation are just emerging. It is likely that many factors, including capillary physical forces, dietary constituents, metabolic injury, and the effects of local growth factors and products of inflammation, converge on a common pathway that generates matrix components within the glomerulus and mesangium. The process to accumulate matrix in response to injury likely parallels the process of scarring and wound healing critical for the survival of other tissue types. In the kidney, however, the generation of matrix may sufficiently disrupt the normal nephron architecture to render it useless. It may be possible in the future to modify the "normal" healing process in the glomerulus so that a loss of functioning nephrons does not hold the remaining glomeruli at increased risk for sclerosis.

REFERENCES

1. Diamond JR, Karnovsky MJ: Focal and segmental glomerulosclerosis: analogies to atherosclerosis. Kidney Int 33:917, 1988
2. Harding MB, Harrison JK, Phillips HR et al: Renal artery stenosis: incidence and associated risk factors in 1303 patients undergoing routine cardiac catheterization. J Am Soc Nephrol 1:334, 1990
3. Kaplan C, Pasternack B, Shah H: Age-related incidence of sclerotic glomeruli in human kidneys. Am J Pathol 80:227, 1975
4. Rowe J, Andres R, Tobin J et al: The effect of age on creatinine clearance in men: a cross-sectional and longitudinal study. J Gerontol 31:155, 1976
5. Brenner BM: Nephron adaptation to renal injury or ablation. Am J Physiol 249:F324, 1985
6. Keane WF, Kasiske BL, O'Donnell MP: Lipids and progressive glomerulosclerosis: a model analogous to atherosclerosis. Am J Nephrol 8:261, 1988
7. Border WA, Okuda S, Nakamura T: Extracellular matrix and glomerular disease. Semin Nephrol 9:307, 1989
8. Veis JH, Yamashita W, Liu YJ, Ooi BS: The biology of mesangial cells in glomerulonephritis. Proc Soc Exp Biol Med 4:160, 1990
9. Lindeman RD, Tobin JD, Shock NW: Association between blood pressure and the rate of decline in renal function with age. Kidney Int 26:861, 1984
10. Bonner G, Shelp WD, Newton M, Rieselbach RE: Factors influencing the increase in glomerular filtration rate in the remaining kidney of transplant donors. Am J Med 55:169, 1973
11. Vincenti F, Amend WJ, Kaysen G et al: Long-term renal function in kidney donors: sustained compensatory hyperfiltration with no adverse effects. Transplantation 36:626, 1983
12. Weiland D, Simmons R, Ascher N, Najarian J: Information on 628 living-related kidney donors at a single institution, with long-term follow-up in 472 cases. Transplant Proc 16:5, 1984
13. Anderson C, Velosa J, Frohnert P et al: The risks of unilateral nephrectomy: status of kidney donors 10 to 20 years postoperatively. Mayo Clin Proc 60:367, 1985
14. Brazy PC: Natural history of renal function in successful renal transplants. Clin Res 39:455A, 1991
15. Kaiske B, Uman AJ: Persistent hyperlipidemia in renal transplant patients. Medicine 66:309, 1987
16. Walser M, Ward L: Progression of chronic renal failure is related to glucocorticoid production. Kidney Int 34:859, 1988
17. Mauer SM, Goetz FC, McHugh LE et al: Long-term study of normal kidneys transplanted into patients with type I diabetes. Diabetes 38:516, 1989
18. Williams PS, Fass G, Bone JM: Renal pathology and proteinuria determine progression in untreated mild/moderate chronic renal failure. Q J Med 67:343, 1988
19. Mitch WE, Walser M, Buffington GA, Lemann J, Jr: A simple method for estimating progression of chronic renal failure. Lancet 2:1326, 1976
20. Brazy PC, Stead WW, Fitzwilliam JF: Progression of renal insufficiency: role of blood pressure. Kidney Int 35:670, 1989
21. Rutherford WE, Blondin J, Miller JP et al: Chronic progressive renal disease: rate of change of serum creatinine concentration. Kidney Int 11:62, 1977

22. Klahr S, Purkerson ML, Heifets M: Factors that may retard the progression of renal disease. Kidney Int, suppl. 22. 32:S-35, 1987
23. Walser M, Drew HH, LaFrance ND: Creatinine measurements often yield false estimates of progression in chronic renal failure. Kidney Int 34:412, 1988
24. Lemann J, Jr, Bidani A, Bain RP et al: Use of the serum creatinine to estimate glomerular filtration rate in health and early diabetic nephropathy. Am J Kidney Dis 16:236, 1990
25. Maschio GL, Oldrizzi L, Tessitore N et al: Effects of dietary protein and phosphorus restriction on the progression of early renal failure. Kidney Int 22:371, 1982
26. Bergstrom J, Alvestrand A, Bucht H, Gutierrez A: Progression of chronic renal failure in man is retarded with more frequent clinical follow-up and better blood pressure control. Clin Nephrol 25:1, 1985
27. Hakim RM, Lazarus JM: Progression of chronic renal failure. Am J Kidney Dis 14:396, 1989
28. Brazy PC, Fitzwilliam JF: Progressive renal disease: role of race and antihypertensive medications. Kidney Int 37:1113, 1990
29. Tierney WM, McDonald CJ, Luft FC: Renal disease in hypertensive adults: effect of race and type II diabetes mellitus. Am J Kidney Dis 13:485, 1989
30. Rostand SG, Brown G, Kirk KA, et al: Renal insufficiency in treated essential hypertension. N Engl J Med 320:684, 1989
31. Whelton PK, Klag MJ: Hypertension as a risk factor for renal disease. Review of clinical and epidemiological evidence. Hypertension 13:I19, 1989
32. McClellan W, Tuttle E, Issa A: Racial differences in the incidence of hypertensive ESRD are not entirely explained by differences in prevalence of hypertension. Am J Kidney Dis 12:285, 1988
33. Moyer JH, Heider C, Pevey K, Ford RV: The effect of treatment on the vascular deterioration associated with hypertension, with particular emphasis on renal function. Am J Med 24:177, 1958
34. Abraham PA, Opsahl JA, Halstenson CE, Keane WF: Efficacy and renal effects of enalapril therapy for hypertensive patients with chronic renal insufficiency. Arch Intern Med 148:2358, 1988
35. Ruilope LM, Miranda B, Morales JM et al: Converting enzyme inhibition in chronic renal failure. Am J Kidney Dis 13:120, 1989
36. Eliahou HE, Cohen D, Hellberg B et al: Effect of the calcium channel blocker nisoldipine on the progression of chronic renal failure in man. Am J Nephrol 8:285, 1988
37. Mitchell HC, Graham RM, Pettinger WA: Renal function during long-term treatment of hypertension with minoxidil: comparison of benign and malignant hypertension. Ann Intern Med 93:676, 1980
38. Humphrey LL, Ballard DJ, Frohnert PP et al: Chronic renal failure in non-insulin-dependent diabetes mellitus. A population-based study in Rochester, Minnesota. Ann Intern Med 111:788, 1989
39. Hasslacher C, Ritz E, Wahl P, Michael C: Similar risks of nephropathy in patients with type I or type II diabetes mellitus. Nephrol Dial Transplant 4:859, 1989
40. Seaquist ER, Goetz FC, Rich S, Barbosa J: Familial clustering of diabetic kidney disease. Evidence for genetic susceptibility to diabetic nephropathy. N Engl J Med 320:1161, 1989
41. Krolewski AS, Canessa M, Warram JH et al: Predisposition to hypertension and

susceptibility to renal disease in insulin-dependent diabetes mellitus. N Engl J Med 318:140, 1988

42. Gall M-A, Rossing P, Jensen JS et al: Red cell Na/Li countertransport in non-insulin-dependent diabetics with diabetic nephropathy. Kidney Int 39:135, 1991
43. Jones SL, Trevisan R, Tariq T et al: Sodium–lithium countertransport in microalbuminuria insulin dependent diabetic patients. Hypertension 15:570, 1990
44. Mogensen CE: Long-term anti-hypertensive treatment inhibiting progression of diabetic nephropathy. Br Med J 285:685, 1982
45. Parving HH, Smidt UM, Andersen AR, Svendsen PAA: Early aggressive antihypertensive treatment reduces rate of decline in kidney function in diabetic nephropathy. Lancet 1:1175, 1983
46. Fogo A, Yoshida Y, Glick A et al: Serial micropuncture analysis of glomerular function in two rat models of glomerular sclerosis. J Clin Invest 82:322, 1988
47. Fogo A, Ickikawa I: Evidence for the central role of glomerular growth promoters in the development of sclerosis. Semin Nephrol 9:329, 1989
48. O'Donnell MP, Kasiske BL, Schmitz PG, Keane WF: High protein intake accelerates glomerulosclerosis independent of effects on glomerular hemodynamics. Kidney Int 37:1263, 1990
49. Mounier F, Foidart J-M, Gubler M-C: Distribution of extracellular matrix glycoproteins during normal development of human kidney. Lab Invest 54:394, 1986
50. Ekblom P, Thesleff I: The role of transferrin and extracellular matrix components in kidney development. Mod Cell Biol 4:85, 1985
51. Ekblom M, Klein G, Mugrauer G et al: Transient and locally restricted expression of laminin A chain mRNA by developing epithelial cells during kidney organogenesis. Cell 60:337, 1990
52. Abrahamson DR, Perry EW: Evidence for splicing new basement membrane into old during glomerular development in newborn rat kidneys. J Cell Biol 103:2489, 1986
53. Laurie GW, Horikoshi S, Killen PD et al: In situ hybridization reveals temporal and spatial changes in cellular expression of mRNA for a laminin receptor, laminin, and basement membrane (type IV) collagen in the developing kidney. J Cell Biol 109:1351, 1989
54. Martin GR, Timpl R, Kuhn K: Basement membrane proteins: molecular structure and function. Adv Protein Chem 39:1, 1988
55. Timpl R, Aumailley M: Biochemistry of basement membranes. Adv Nephrol 18:59, 1989
56. Gehron-Roby P, Martin GR: Type IV collagen contains two distinct chains in separate molecules. Coll Rel Res 1:27, 1981
57. Butkowski RJ, Wieslander J, Kleppel M et al: Basement membrane collagen in the kidney: regional localization of novel chains related to collagen IV. Kidney Int 35:1195, 1989
58. Saus J, Wieslander J, Langeveld JPM et al: Identification of the Goodpasture antigen as the α3(IV) chain of collagen IV. J Biol Chem 263:13374, 1988
59. Morrison DE, Germino GG, Reeders ST: Use of the polymerase chain reaction to clone and sequence a cDNA encoding the bovine α3 chain of type IV collagen. J Biol Chem 266:34, 1991
60. Gunwar S, Saus J, Noelken ME, Hudson BG: Glomerular basement membrane: identification of fourth chain, α4, of type IV collagen. J Biol Chem 265:5466, 1990

61. Hostikka SL, Eddy RL, Byers MS et al: Identification of a distinct type IV collagen α chain with restricted kidney distribution and assignment of its gene to the locus of X chromosome-linked Alport syndrome. Proc Natl Acad Sci USA 87:1606, 1990
62. Kleppel MM, Kashtan C, Santi PA et al: Distribution of familial nephritis antigen in normal tissue and renal basement membranes of patients with homozygous and heterozygous Alport familial nephritis. Lab Invest 61:278, 1989
63. Barker DF, Hostikka SL, Zhou J et al: Identification of mutations in the COL4α5 collagen gene in Alport syndrome. Science 248:1224, 1990
64. Kleppel MM, Michael AF: Expression of novel basement membrane components in the developing human kidney and eye. Ann J Anat 187:165, 1990
65. Houser MT, Scheinman JI, Basgen J et al: Preservation of mesangium and immunohistochemically defined antigens in glomerular basement membrane isolated by detergent extraction. J Clin Invest 69:1169, 1982
66. Martinez-Hernandez A, Gay S, Miller EJ: Ultrastructural localization of type V collagen in rat kidney. J Cell Biol 92:343, 1982
67. Rosenblum ND, Karnovsky MJ, Olsen BR: Non-fibrillar collagens synthesized by rat mesangial cells. J Am Soc Nephrol 1:785, 1990
68. Beck K, Hunter I, Engel J: Structure and function of laminin: anatomy of a multidomain glycoprotein. FASEB J 4:148, 1990
69. Kleinman HK, Ebihara I, Killen PD et al: Genes for basement membrane proteins are coordinately expressed in differentiating F9 cells but not in normal adult murine tissues. Dev Biol 122:373, 1987
70. Boot-Handford RP, Kurkinen M, Prockop DJ: Steady-state levels of mRNAs coding for the type IV collagen and laminin polypeptide chains of basement membranes exhibit marked tissue-specific stoichiometric variations in the rat. J Biol Chem 262:12475, 1987
71. Sanes JR, Engvall E, Butkowski R et al: Molecular heterogeneity f basal laminae: isoforms of laminin and collagen IV at the neuromuscular junction and elsewhere. J Cell Biol 111:1685, 1990
72. Paulsson M: The role of Ca^{2+} binding in the self-aggregation of laminin-nidogen complexes. J Biol Chem 263:5425, 1988
73. Bender BL, Jaffe R, Carlin B, Chung AE: Immunolocalization of entactin, a sulfated basement membrane component in rodent tissues, and comparison with GP-2 (laminin). Am J Pathol 103:419, 1981
74. Hassell JR, Leyshon WC, Ledbetter SR et al: Isolation of two forms of basement membrane proteoglycans. J Biol Chem 260:8098, 1985
75. Klein DJ, Brown DM, Oegema PR et al: Glomerular basement membrane proteoglycans are derived from a large precursor. J Cell Biol 106:963, 1988
76. Noonan DM, Horigan EA, Ledbetter SR et al: Identification of cDNA clones encoding different domains of the basement membrane heparan sulfate proteoglycans. J Biol Chem 263:16379, 1988
77. Kemeny E, Fillit HM, Damle S et al: Monoclonal antibodies to heparan sulfate proteoglycan: development and application to the study of normal tissue in kidney biopsies. Connect Tissue Res 18:9, 1988
78. Oberley TD, Murphy-Ulrich JE: Fibronectin and the kidney. p. 309. In Mosher DF (ed): Fibronectin. Academic Press, San Diego, 1989
79. Wight TN, Raugi GJ, Mumby SM, Bornstein P: Light microscopic immunolocation of thrombospondin in human tissues. J Histochem Cytochem 33:295, 1985

80. McCarthy KJ, Accavitti MA, Couchman JR: Immunologic characterization of a basement membrane-specific chondroitin sulfate proteoglycan. J Cell Biol 109:3187, 1989
81. Bianco P, Fisher LW, Young MF et al: Expression and localization of the two small proteoglycans biglycan and decorin in developing human skeletal and non-skeletal tissues. J Histochem Cytochem 38:1549, 1990
82. Rennke HG, Klein PS: Pathogenesis and significance of nonprimary focal and segmental glomerulosclerosis. Am J Kidney Dis 6:443, 1989
83. Foidart JM, Foidart JB, Mahieu PR: Synthesis of collagen and fibronectin by glomerular cells in culture. Renal Physiol 3:183, 1980
84. Morel-Maroger L, Killen PD, Chi E, Striker GE: The composition of glomerulosclerosis I. Studies in focal sclerosis, crescentic glomerulonephritis, and membranoproliferative glomerulonephritis. Lab Invest 5:181, 1984
85. Oomura A, Nakamura T, Arakawa M et al: Alterations in the extracellular matrix components in human glomerular diseases. Virchows Arch [A] 415:151, 1989
86. Yoshioka K, Takemura T, Tohda M et al: Glomerular localization of type III collagen in human kidney disease. Kidney Int 35:1203, 1989
87. Adler S, Striker LJ, Striker GE et al: Studies of progressive glomerular sclerosis in the rat. Am J Pathol 123:553, 1986
88. MacKay K, Striker LJ, Pinkert CA et al: Glomerulosclerosis and renal cysts in mice transgenic for the early region of SV40. Kidney Int 32:827, 1987
89. Doi T, Striker L, Quaife C et al: Progressive glomerulosclerosis develops in transgenic mice chronically expressing growth hormone and growth hormone releasing factor but not in those expressing insulinlike growth factor I. Am J Pathol 131:398, 1988
90. Quaife CJ, Mathews LS, Pinkert CA et al: Histopathology associated with elevated levels of growth hormone and insulin-like growth factor I in transgenic mice. Endocrinology 124:40, 1989
91. Adler SA, Kopp JB, Dickie P et al: Enhanced extracellular matrix accumulation in mice transgenic for HIV genes. J Am Soc Nephrol 1:513, 1990
92. Okuda S, Languino LR, Ruoslahti E, Border WA: Elevated expression of transforming growth factor-b and proteoglycan production in experimental glomerulonephritis. J Clin Invest 86:453, 1990
93. Klein DJ, Brown DM, Kim Y, Oegema TR: Proteoglycans synthesized by human glomerular mesangial cells in culture. J Biol Chem 265:9533, 1990
94. Hasslacher C, Reichenbacher R, Gechter F, Timpl R: Glomerular basement membrane synthesis and serum concentration of type IV collagen in streptozotocin-diabetic rats. Diabetologia 26:150, 1984
95. Klein DJ, Brown DM, Oegema TR: Glomerular proteoglycan in diabetes. Diabetes 35:1130, 1986
96. Shimonura H, Spiro RG: Studies on the macromolecular components of human glomerular basement membrane and alterations in diabetes. Diabetes 36:374, 1987
97. Falk RJ, Scheinman JI, Mauer SM, Michael AF: Polyantigentic expansion of basement membrane constituents in diabetic nephropathy. Diabetes suppl. 2 32:34, 1983
98. Bruneval P, Foidart JM, Nochy D et al: Glomerular matrix proteins in nodular glomerulosclerosis in association with light chain deposition disease and diabetes mellitus. Hum Pathol 16:477, 1985

99. Karttunen T, Risteli J, Autio-Harmainen H, Risteli L: Effect of age and diabetes on type IV collagen and laminin in human kidney cortex. Kidney Int 30:586, 1986
100. Abrass CK, Peterson CV, Rauji GJ: Phenotypic expression of collagen types in mesangial matrix of diabetic and non-diabetic rats. Diabetes 37:1695, 1988
101. Kopp JB, Bodkin N, Marinos NJ et al: Increased laminin in nodular glomerulosclerosis affecting rhesus monkeys with non-insulin dependent diabetes mellitus J Am Soc Nephrol 1:551, 1990
102. Kim Y, Kleppel MM, Butkowski R et al: Differential expression of basement membrane collagen chains in diabetic nephropathy. Am J Pathol 138:413, 1991
103. Killen PD, Ebihara I, Martin GR et al: mRNA levels for laminin and collagen IV chains are elevated in diabetic kidneys. Kidney Int 31:171, 1987
104. Paulsom R, Kurkinem M, Prockop DJ, Boot-Handford RP: Increased steady-state levels of laminin B1 mRNA in kidneys of long-term streptozotocin-diabetic rats. J Biol Chem 263:10072, 1988
105. Ledbetter S, Copeland EJ, Noonan D et al: Altered steady-state mRNA levels of basement proteins in diabetic mouse kidneys and thromboxane synthetase inhibitors. Diabetes 39:196, 1990
106. Burbelo PD, Klotman P, Bruggeman L et al: Regulation of basement membrane genes. Critical Reviews in Eukaryotic Gene Expression 1:1, 1990
107. Bruggeman LA, Burbelo PD, Yamada Y et al: Transcriptional factors for type IV collagen gene expression identified in nuclear extracts from mouse and human kidney. Kidney Int 37:218, 1990
108. Mene P, Simonson MS, Dunn MJ: Physiology of the mesangial cell. Physiol Rev 69:1347, 1989
109. Craven PA, Caines MA, DeRubertis FR: Sequential alterations in glomerular prostaglandin and thromboxane synthesis in diabetic rats: relationship to hyperfiltration in early diabetes. Metabolism 36:95, 1987
110. Purkerson ML, Joist JH, Yates J et al: Inhibition of thromboxane synthesis ameliorates the progressive kidney disease of rats with subtotal renal ablation. Proc Natl Acad Sci USA 82:193, 1985
111. Zoja C, Perico N, Corna D et al: Thromboxane synthesis inhibition increases renal prostacyclin and prevents renal disease progression in rats with remnant kidney. J Am Soc Nephrol 1:799, 1990
112. Klahr S, Heifets M, Purkerson ML: The influence of anti-coagulation on the progression of experimental renal disease. p. 45. In Mitch WE, Brenner BM, Stein JH (eds): The Progressive Nature of Renal Disease. Churchill Livingstone, New York, 1986
113. Rohrbach DH, Hassell JR, Kleinman HK, Martin GR: Alterations in the basement membrane (heparan sulfate) proteoglycan in diabetic mice. Diabetes 31:185, 1982
114. Cagliero E, Maiello M, Boeri D et al: Increased expression of basement membrane components in human endothelial cells cultured in high glucose. J Clin Invest 82:735, 1988
115. Ayo SH, Radnik RA, Glass WF et al: Increased fibronectin, laminin and type IV collagen synthesis and expression of mRNA in renal glomerular mesangial cells grown in medium containing high glucose. Am J Physiol 260:F185, 1991

3

Lipid Abnormalities after Renal Transplantation

Connie L. Manske
Bertram L. Kasiske

INTRODUCTION

Although hyperlipidemia is a well-recognized complication of renal transplantation,[1–17] the clinical importance of lipid abnormalities in renal transplant recipients has not been well defined. In the general population, increased low-density lipoprotein (LDL) and decreased high-density lipoprotein (HDL) are recognized risk factors for atherosclerosis, and lowering LDL reduces the risk of coronary artery disease.[18–24] In the transplant population, limited data exist on the relationship between coronary artery disease and hyperlipidemia. However, cardiovascular disease is currently a leading cause of death in renal transplant recipients and is likely to increase as patient and graft survival continue to improve.[25–30] Therefore, an understanding of post-transplant lipid abnormalities and their potential role in accelerated atherogenesis is of great importance. This chapter will summarize what is known about atherosclerotic vascular disease and lipid metabolism after renal transplantation. Management of post-transplant hyperlipidemia will be discussed, based on studies of the general population as well as the very limited data available in renal transplant recipients.

OVERVIEW OF LIPOPROTEIN METABOLISM

Lipoproteins consist of an outer monolayer of free cholesterol, phospholipids, and apolipoproteins surrounding a disorganized core of hydrophobic cholesteryl esters and triglycerides. The lipoproteins, along with the enzymes lipoprotein lipase and lecithin cholesterol acyltransferase (LCAT), are part of a complex system that delivers cholesterol and triglyceride to peripheral tissues and returns cholesterol to the liver for metabolism and excretion as bile acids.[31] The major classes of lipoproteins are summarized in Table 3-1. As lipoproteins travel through the circulation, surface apoproteins are recognized by cell surface lipases and lipoprotein receptors, resulting in transfer of triglyceride and cholesterol ester to tissues that require them.

Lipoprotein transport can be broken down into three major pathways: (1) dietary fat and cholesterol are transported to the liver and peripheral tissues via lipoproteins of the exogenous pathway (Fig. 3-1), (2) cholesterol and triglycerides reincorporated into lipoproteins in the liver are transported to the periphery via the endogenous pathway (Fig. 3-2), and (3) reverse cholesterol transport is hypothesized to occur via the HDL particle, recycling cholesterol in the periphery back to the liver (Fig. 3-3).[32] In addition, there is substantial transfer of triglycerides and cholesteryl esters between plasma lipoproteins via the action of lipid transfer proteins.[33]

In the exogenous pathway, chylomicrons are formed from intestinal free fatty acids and esterified cholesterol and combined with apoprotein (apo) B_{48}. After traversing the thoracic duct, the chylomicron acquires apoproteins C-II and E in the blood stream. The presence of apo C-II and apo E allows chylomicrons to be recognized by lipoprotein lipase, an enzyme bound to the endothelium of capillaries in adipose tissue and muscle. Lipoprotein lipase, which requires apo C-II as a cofactor, hydrolyzes chylomicron and very low-density lipoprotein (VLDL) core triglycerides to monoglycerides, free fatty acids, and remnant particles (intermediate-density lipoproteins [IDL]). The resultant fatty acids are taken up in tissues where they are stored or used for fuel. As triglycerides are removed and transported to tissues, the

Table 3-1. Effect of Diet on Lipids in Renal Transplant Recipients

Study	No. of Patients	Study Length (weeks)	Immunosuppression	Diet	Percent Obese	Compliance	Pre-, Post-Transplant	Results[a] Weight (kg)	Cholesterol (mg/dl)	Triglyceride (mg/dl)
Ponticelli et al.[116] (1975)	6	4	4/17 patients:CSA	Hypocaloric 43% CHO 23% protein 33% fat	100	NA	Pre Post	67 ± 3 65 ± 3	243 ± 25 216 ± 20	327 ± 39 160 ± 7[b]
Disler et al.[117] (1981)	21	52	AZA	Hypocaloric 50% CHO 26% protein 25% fat P:S > 1 <300 mg cholesterol	100	62%	Pre Post	129 ± 5 (% IBW) 121 ± 5	355 ± 19 293 ± 19[b]	336 ± 53 283 ± 44
Shen et al.[96] (1983)	12	13	AZA	Hypocaloric < 50% CHO < 35% fat P:S > 1	75	Kcal, 75% Cholesterol, 100% Fat, 42% P:S, 42%	Pre Post	NA	313 ± 19 248 ± 12[b]	321 ± 63 198 ± 30[b]
Nelson et al.[97] (1988)	9	2	CSA	Hypocaloric 50% CHO 21% protein 29% fat P:S > 1 < 300 mg cholesterol	100	100%	Pre Post	79 ± 4 76 ± 4[b]	316 ± 36 266 ± 30[b]	423 ± 94 295 ± 60[b]
Moore et al.[118] (1990)	17	8	AZA	Isocaloric 50–60% CHO 10–20% protein < 30% fat < 300 mg cholesterol	76	Fat, 46% Cholesterol, 70%	Pre Post	76 ± 3 75 ± 3[b]	262 ± 14 241 ± 13[b]	179 ± 25 169 ± 17

Abbreviations: P:S, ratio of polyunsaturated fat to saturated fat; CSA, cyclosporine; AZA, azathiaprine; IBW, ideal body weight; CHO, carbohydrate; NA, not available.

[a] Results expressed as mean ± SEM

[b] Significantly different from baseline ($p < .05$)

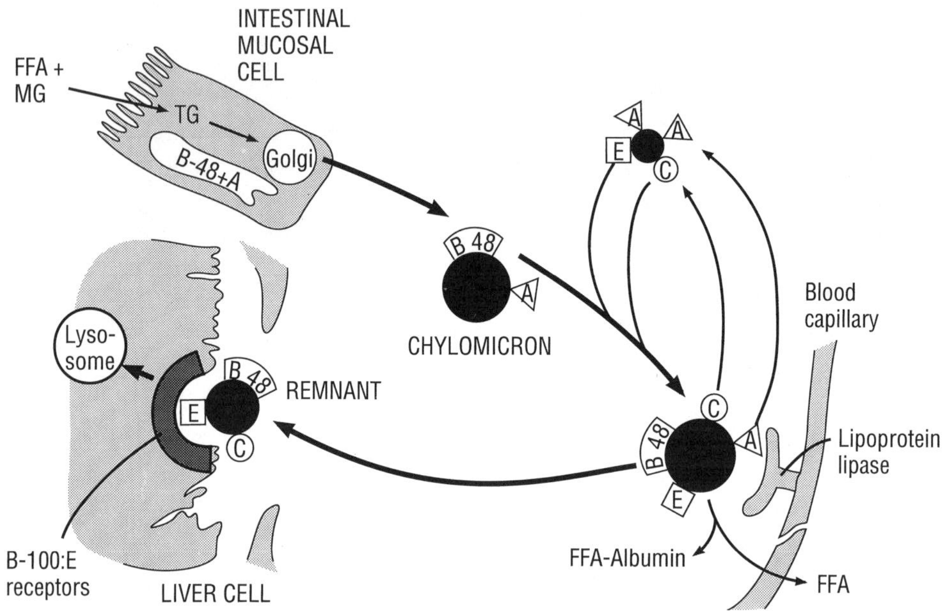

Fig. 3-1. The exogenous pathway for transport of dietary lipids to the liver and periphery. FFA, free fatty acid; MG, monoglycerides; TG, triglycerides; apoproteins: A, C, E, B_{48}, B_{100}. (From Havel,[149] with permission.)

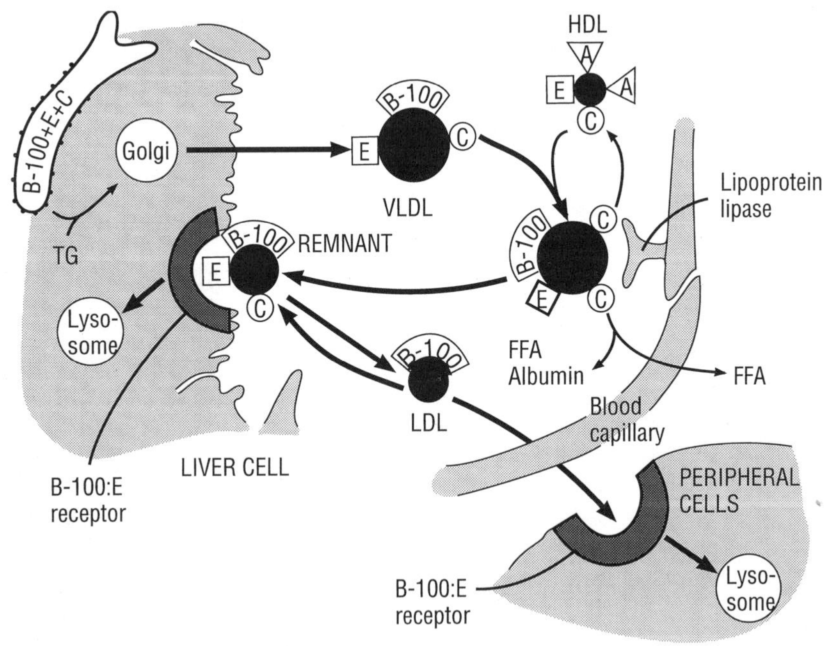

Fig. 3-2. The endogenous pathway for transport of hepatically synthesized lipids to the periphery. HDL, high-density lipoprotein; LDL, low-density lipoprotein; other abbreviations as in Fig. 3-1. (From Havel,[149] with permission.)

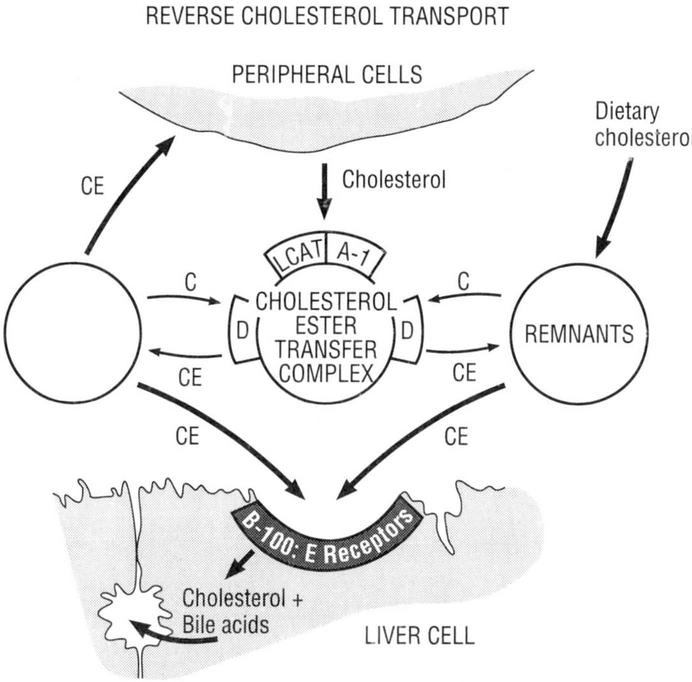

Fig. 3-3. The HDL cholesteryl-ester transfer complex, the major pathway by which cholesterol is transported to the liver for excretion. CE, cholesteryl esters; C, cholesterol; LDL, low-density lipoprotein; LCAT, lecithin-cholesterol acyltransferase; apoproteins: A_1 and D. (From Havel,[149] with permission.)

surface lipids and apoproteins transfer to HDL particles. Eventually, the loss of apo C-II during the process decreases the affinity of the chylomicron for lipoprotein lipase. The IDL formed from chylomicron metabolism is rapidly taken up by hepatic parenchymal cells and broken down, possibly via a chylomicron remnant receptor that recognizes apo E. Cholesterol ester from the IDL is catabolized to free cholesterol and then secreted into bile or reincorporated into lipids. The transport of lipids via the exogenous pathway is very rapid, so that chylomicrons and chylomicron remnants are not usually present after an overnight fast and therefore do not normally contribute to fasting cholesterol or triglyceride levels.

In the endogenous pathway, VLDLs are formed in the liver in the fasted state from endogenous triglyceride and cholesterol in combination with apo B_{100}, apo E, and apo C-II. In the circulation, VLDL triglycerides are catabolized by lipoprotein lipase. The resulting IDL has one of two fates: removal directly by the liver via the LDL receptor in a manner analogous to the chylomicron remnant, or further degradation to LDL by removal of the remaining triglycerides.

LDL, the major cholesterol-carrying lipoprotein in the circulation, has a core consisting almost entirely of cholesteryl esters. Approximately 75 percent of circulating LDL is cleared via LDL receptors in the liver or in cholesterol-requiring tissue, such as the gonads, adrenals, and rapidly dividing cells. The remainder is taken up in the reticuloendothelial system by non–receptor-mediated mechanisms and may be instrumental in atherogenesis because of the participation of vascular smooth muscle cells and macrophages in this process.

The LDL receptor plays a prominent role in lipoprotein metabolism.[34] LDL receptors are present in all cells, but are particularly abundant on hepatic parenchymal cells. When receptor number is high, LDL can be efficiently taken up and metabolized by the liver. Receptor number is regulated by cell content of cholesterol; downregulation occurs when cholesterol content is high. If receptor number is low, a greater proportion of LDL will be available for non–receptor-mediated catabolism via smooth muscle cells, macrophages, and reticuloendothelial cells.

Three basic mechanisms can cause high serum LDL cholesterol: defective LDL clearance, LDL overproduction, and overloading of LDL particles with cholesterol ester. Delayed receptor-mediated clearance of LDL can occur as a result of an abnormality in LDL receptor or a defective LDL that binds poorly to the receptor.[31] Both defects have been documented in humans. Overproduction of LDL may result from overproduction of apo B-containing lipoproteins by the liver or from decreased hepatic uptake of IDL resulting in an increased intravascular conversion of IDL to LDL. If LDL becomes abnormally enriched with cholesterol esters, LDL cholesterol will increase even if the number of LDL particles in the circulation does not increase.

The role of HDL and the reverse cholesterol transport pathway in lipoprotein metabolism is uncertain, since regulation of HDL synthesis and catabolism is poorly understood.[32] Unlike chylomicrons and VLDL, HDL is not secreted in near-final form but as a simple bilayer of phospholipids surrounded by apo E. HDL particles can be thought of as a biproduct of triglyceride metabolism, since they are largely composed of redundant surface lipids and apoproteins generated during the catabolism of triglyceride-rich lipoproteins by lipoprotein lipase.[35] As might be expected, HDL levels are positively correlated with lipoprotein lipase activity and inversely correlated with VLDL levels. Therefore, high HDL levels may indicate efficient triglyceride metabolism.[36,37] HDL exists as part of a cholesterol ester transfer complex containing LCAT, apo A-I (a cofactor for the enzyme), and apo D (Fig. 3-3). Free cholesterol is avidly taken up by HDL and esterified by LCAT, creating a free cholesterol gradient from cells to the HDL transfer complex. The final step in reverse cholesterol transport is transfer of cholesteryl esters back to VLDL and LDL to be returned to the liver for catabolism. HDL reverse cholesterol transport may be the major pathway by which free cholesterol can be removed from cells (e.g., atherosclerotic plaques), and may explain why high levels of HDL are associated with a reduced risk for cardiovascular disease.

Although discrete pathways of lipoprotein metabolism have been constructed, it is likely that a dynamic equilibrium exists among the lipoprotein subclasses. Transfer between plasma lipoproteins occurs via the action of lipoprotein transfer proteins and is driven by the difference in effective concentration of cholesterol esters and triglycerides in different lipoproteins in plasma.[33] In general, lipolysis mediated by lipoprotein lipase promotes exchange of lipids between lipoproteins.[37] VLDL and LDL play major roles as acceptors of cholesteryl ester generated in HDL particles. If abnormalities in VLDL or LDL composition occur, transfer is inhibited and cholesterol return to the liver for catabolism or excretion is impaired. Inhibition of transfer has been shown to be associated with atherogenesis.[38]

LIPOPROTEINS AND ATHEROGENESIS

An atheroma consists of lipid-laden macrophages, smooth muscle cells, connective tissue matrix, and cholesterol crystals. Two theories have evolved to explain the origin of the atherosclerotic plaque.[39] The endothelial injury hypothesis proposes that endothelial injury occurs, causing platelet aggregation. This presumably triggers the release of mediators and growth factors, leading to smooth muscle proliferation and the secretion of connective tissue matrix elements, with lipid accumulation occurring secondarily. The lipid infiltration hypothesis proposes that hypercholesterolemia is the major cause of atherosclerosis. This is based on the knowledge that accumulation of cholesterol in the arterial wall is a hallmark of atherosclerosis and that atherosclerosis can be produced experimentally by increasing plasma cholesterol levels in animals. It is likely that both theories are correct. Indeed, lipoproteins have been shown to be directly injurious to endothelial cells and to activate platelets.[40,41]

There is clear epidemiologic evidence that hypercholesterolemia is a necessary part of atherogenesis. For example, populations with low LDL cholesterol levels (such as the Japanese) have a low incidence of coronary heart disease, even though the prevalence of both hypertension and smoking is high. Furthermore, modification of LDL cholesterol levels has been shown to be an effective way to decrease coronary heart disease events.[18,20–23] In some cases, lowering serum cholesterol has resulted in regression of established atherosclerotic plaques.[42,43]

ATHEROSCLEROTIC VASCULAR DISEASE IN RENAL TRANSPLANT RECIPIENTS

Atherosclerosis is a major cause of morbidity and mortality after renal transplantation, particularly in diabetic patients. Prospective series have revealed that 25 to 42 percent of diabetic transplant candidates have significant coronary artery disease (at least one vessel with greater than 70 percent stenosis) prior to transplantation.[44–47] Analysis of 237 diabetic patients transplanted at the University of Minnesota between 1980 and 1985 revealed that 18 patients died from documented myocardial infarction or sudden death within the first 2 years after transplantation (Manske CL: unpublished observation). It is clear that a high proportion of diabetic patients have established coronary artery disease at the time of transplantation. In nondiabetic patients undergoing renal transplantation, no prospective coronary angiography data are available to evaluate the prevalence of coronary artery disease. However, two early series containing few diabetic patients have documented an 11- to 25-fold increase in myocardial infarction within the first 3 years after renal transplant compared with age-matched controls from the general population, suggesting that significant atherosclerotic vascular disease was present prior to transplantation in a large proportion of nondiabetic renal transplant recipients.[25,48]

The prevalence of known risk factors for coronary artery disease increases when patients develop chronic renal failure. The majority of patients develop hypertension, particularly if the underlying renal disease is diabetes.[49–53] Lipid abnormalities regularly seen in chronic renal failure include low HDL cholesterol and hypertriglyceridemia, presumably as a result of inhibition of lipoprotein lipase.[10,54–62] If significant proteinuria is present, cholesterol increases; this has been correlated with the magnitude of proteinuria.[63–66] Insulin resistance, a known correlate of coronary artery disease in the general population,[67,68] is frequently present in chronic renal failure.[69,70] Other aspects of the azotemic milieu, such as hyperparathyroidism and the accumulation of growth factors, may also play a role in accelerating atherogenesis.[71]

In diabetic patients, epidemiologic data confirm that chronic renal failure is a modifier of atherosclerosis and accelerates the progression of early atherosclerotic lesions to severe coronary artery disease. Of 292 diabetic patients followed at the Joslin Clinic for 20 to 40 years, 35 percent had died by age 55 of coronary artery disease; diabetic patients with chronic renal failure were 15 times more likely to die than diabetic patients with normal renal function.[72] The role of chronic renal failure as a potent accelerator of coronary artery disease in diabetic patients was corroborated by data from the Steno Memorial Clinic, where clinical coronary artery disease was present in 40 percent of azotemic diabetic patients within 6 years of proteinuria onset.[49]

Although the increased prevalence of established coronary artery disease at the time of transplantation is well documented, whether atherogenesis is accelerated in the post-transplant period is controversial. One early series that did not include diabetic patients found no increase in incidence in coronary artery disease in 68 long-term survivors (5 to 13 years) of renal transplantation.[73] However, in several large series, cardiovascular disease was the major cause of death of long-term transplant survivors (6 to 20 years) regardless of donor source.[25,27–29] The prevalence of diabetes in these studies was not given. In a large study of HLA-identical graft recipients, coronary artery disease was the major cause of death only in diabetic transplant recipients.[30] However, in a study of 464 transplant recipients, less than 25 percent of whom were diabetic, the incidence of post-transplant vascular disease was high. Moreover, the increased incidence of vascular disease was evident even in patients with no clinically apparent vascular disease present at the time of transplantation.[74] Altogether, these data suggest that atherosclerotic vascular disease continues to be a significant problem in the late post-transplant period (Fig. 3-4).

HYPERLIPIDEMIA IN TRANSPLANT RECIPIENTS

Hypercholesterolemia

Numerous series have documented the high incidence of hyperlipidemia after renal transplantation. Total serum cholesterol increases significantly after renal transplant in a majority of patients.[8] In one study of 175 patients, 38 percent of nondiabetic and 55 percent of diabetic adults had a cholesterol level greater than 2 SD above normal when measured at least 4 months post-transplant.[5] In a combined series of diabetic and nondiabetic patients with stable renal function for at least 1 year after transplant, cholesterol levels were greater than the 95th percentile of the normal population in 30 percent of the patients.[14] These series may have underestimated the incidence of hypercholesterolemia, since patients who suffered myocardial death in the early post-transplant period were not included.

Several studies have shown a positive correlation between serum cholesterol level and daily, or cumulative, prednisone dose.[4,5,8] Moreover, when prednisone has been withdrawn completely, serum cholesterol declines significantly.[75] Although sometimes attributed to proteinuria or renal dysfunction, the correlation between cumulative prednisone dose and serum cholesterol remained significant when heart transplant recipients with normal renal function have been evaluated.[76] A similar increase in serum choles-

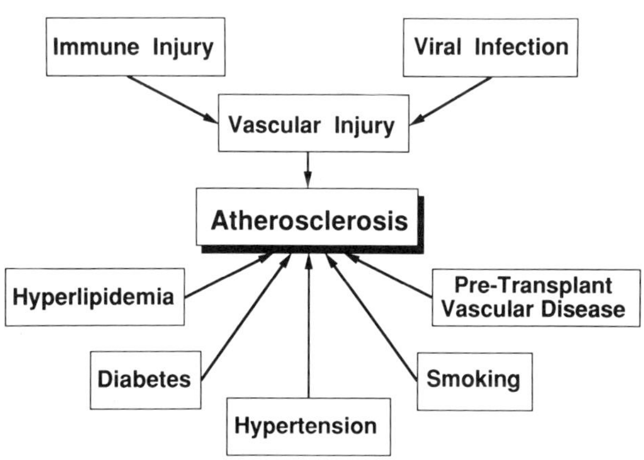

Fig. 3-4. Factors possibly important in the pathogenesis of post-transplant atherosclerosis.

terol was seen in endogenous hypercortisolism (Cushing's syndrome),[77] as well as in other disease states in which exogenous corticosteroids were used.[78] Prednisone may increase serum cholesterol by stimulating hepatic production of lipoproteins. Studies in animals and humans suggest that exogenous corticosteroids augment insulin action, increasing hepatic synthesis of VLDL.[7,77,79] Corticosteroids also stimulate adipocyte hormone-sensitive lipase, thus releasing stored triglycerides as free fatty acids.[80] The availability of additional free fatty acids for hepatic triglyceride synthesis further increases hepatic VLDL synthesis. Because the insulin-potentiating action of corticosteroids increases activity of lipoprotein lipase, increased conversion of VLDL to LDL may lead to an elevation of serum cholesterol. Corticosteroids may also cause hypercholesterolemia by interfering with LDL receptor function.[81]

When cyclosporine was introduced into the field of transplantation, it was hoped that the resulting reduction in prednisone dose would decrease the magnitude of hypercholesterolemia in renal transplant recipients. Unfortunately, cyclosporine was found to cause an increase in serum cholesterol that was independent of its known suppression of glomerular filtration.[12,15–17] Since initial data were obtained in renal transplant recipients, it was difficult to separate the effects of weight gain, antihypertensive drugs, and allograft dysfunction from the effects of cyclosporine on serum cholesterol metabolism. However, a prospective, double-blind, randomized study of 18 patients receiving cyclosporine for amyotrophic lateral sclerosis revealed a 21 percent increase in total cholesterol, a 31 percent increase in LDL cholesterol, and a 12 percent increase in apo B in the cyclosporine-treated patients.[82] In another study, 12 patients randomized to cyclosporine for treatment of psoriasis had a 19 percent increase in total cholesterol compared with controls.[83]

Two possible mechanisms may account for cyclosporine-induced hypercholesterolemia. Cyclosporine is a lipophilic endecapeptide that is transported in the blood in association with lipoproteins.[84] Because of its association with lipoproteins, it is hypothesized that cyclosporine may enter cells via the LDL receptor.[85] Presumably, hypercholesterolemia might occur because of an abnormal interaction between LDL and the LDL receptor secondary to the presence of cyclosporine in the LDL particle.[86] A second mechanism for cyclosporine-induced hypercholesterolemia may relate to its effect on bile acid synthesis. Cyclosporine has been shown to reduce bile acid synthesis by cultured hepatocytes.[87] Cyclosporine may inhibit 26-hydroxylase, an important enzyme in bile acid synthesis. This could reduce the excretion of free cholesterol in bile and result in systemic hypercholesterolemia.

Regardless of the medication used for immunosuppression, hypercholesterolemia and increased LDL are associated with reduced allograft function.[11,13,14,17] Surprisingly, few investigators have examined the influence of proteinuria on hyperlipidemia in transplant recipients.[88,89] Since chronic renal failure per se has not been associated with increased LDL,[90] it is likely that the proteinuria commonly seen in chronic rejection contributes to the association between allograft dysfunction and hypercholesterolemia. Two studies that included proteinuria data have corroborated this hypothesis.[13,14] Patients with reduced renal function were more likely to receive greater amounts of immunosuppressive medications; however, proteinuria continued to be associated with post-transplant hypercholesterolemia when this was taken into account.[14] The mechanism of proteinuria-induced hypercholesteremia in renal transplant recipients has not been studied; however, in nephrotic syndrome, it has been attributed to hepatic overproduction of apo B.[91,92] Independently of the effects of immunosuppressive drugs and proteinuria, the magnitude of post-transplant hypercholesterolemia has been correlated with pretransplant lipid values and with age.[13,14,93]

In summary, hypercholesterolemia after renal transplantation can be thought of as the exaggerated response of a susceptible individual to the defined stresses on cholesterol metabolism of proteinuria, corticosteroids, and cyclosporine (Fig. 3-5). Although cholesterol metabolism has not been studied in transplant recipients, one might expect to find enhanced VLDL and LDL synthesis secondary to corticosteroid-induced free fatty acid overproduction coupled with decreased cholesterol excretion secondary to cyclosporine effects on bile acid metabolism. In addition, LDL receptor function might be hindered by cyclosporine incorporation into LDL particles. In the presence of proteinuria, LDL production would be further enhanced by hepatic overproduction of apo B. The combination of these factors would result in hypercholesteremia in many transplant patients, with exaggerated hypercholesteremia in older patients with pre-existing abnormalities in lipid metabolism.

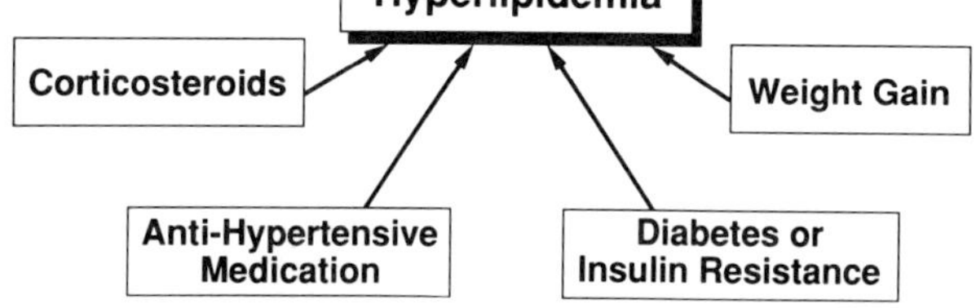

Fig. 3-5. Factors possibly important in the pathogenesis of post-transplant hyperlipidemia.

Hypertriglyceridemia

Hypertriglyceridemia has been reported in 15 to 50 percent of patients after renal transplantation, with the highest prevalence noted in early series when higher doses of corticosteroids were used.[1,3,5,6] Corticosteroid therapy was linked to hypertriglyceridemia in most, but not all,[17] series of transplant recipients. Triglyceride levels correlated with both daily[1,2,5,7,13] and cumulative[4,10] prednisone dose. Corticosteroids are associated with increased peripheral free fatty acid production and increased hepatic VLDL synthesis, as described previously.

The impact of cyclosporine on triglyceride levels is less clear. In one series of 490 patients, the incidence of hypertriglyceridemia was only 15 percent (i.e., much lower than the incidence in early precyclosporine series[17]). However, this may reflect the steroid-sparing effect of cyclosporine. In the only series in which fasting levels were measured in patients without renal disease, 11 patients receiving cyclosporine for psoriasis had a statistically significant increase in serum triglyceride levels after 4 weeks of cyclosporine therapy.[83] This may be secondary to the insulin resistance induced by cyclosporine[94,95] or to its effect on renal function (see below).

Another correlate of hypertriglyceridemia in renal transplant recipients is relative body weight.[5,10,14,17] In one survey, the average transplant recipient's weight was 113 percent of ideal body weight by the first year after transplantation.[17] Increased relative body weight was one of the strongest predictors of hypertriglyceridemia in a large series of patients immunosuppressed with cyclosporine.[17] Intentional weight reduction in transplant recipients was associated with a decrease in triglyceride levels in most series.[6,14,96,97]

The superimposition of chronic renal failure is likely to cause a worsening of hypertriglyceridemia in susceptible patients. Many series of renal transplant recipients have demonstrated a positive correlation between triglyceride level and creatinine.[5,8,9,14,17] In non-transplanted patients with chronic renal failure, hypertriglyceridemia was present in 33 to 50 percent of patients[98] and appeared early in the course of chronic renal failure.[99] However, prospective data is lacking. Production rates of triglycerides were normal in uremia, but catabolism was reduced and remnant particles accumulate.[100] Several studies have shown impairment of lipoprotein lipase activity in patients with chronic renal failure.[101,102] The activity of lipoprotein lipase correlated positively with serum HDL levels and inversely with serum triglyceride levels. The mechanism of azotemic suppression of lipoprotein lipase is not clear, but recent data suggest that hyperparathyroidism plays an important role in suppressing lipoprotein lipase activity.[103,104] The net result is hypertriglyceridemia secondary to slowed metabolism of triglyceride-rich lipoproteins.

Another potential cause of hypertriglyceridemia in renal transplant recipients is the presence of poorly controlled diabetes. As the number of diabetic patients undergoing renal transplantation increases, it is possible that the incidence of post-transplant hypertriglyceridemia will increase. Type I diabetic patients develop hypertriglyceridemia secondary to relative lack of insulin.[105] When blood sugar control is not optimal, free fatty acids are mobilized from the periphery and synthesized into VLDL. Clearance of VLDL is impaired, presumably because lipoprotein lipase requires insulin for activity. Rigorous insulin therapy can result in increased lipoprotein lipase activity and subnormal levels of VLDL.[106] However, many diabetic renal transplant recipients have impaired ability to recognize hyperglycemia and do not maintain tight control. Non-insulin-dependent (type II) diabetic patients probably have hypertriglyceridemia because of insulin resistance. Type II diabetic patients have multiple alterations in VLDL metabolism, including overproduction and a decreased fractional catabolic rate of VLDL triglycerides and apo B.[105] Euglycemic clamp studies have revealed a significant correlation between VLDL triglycerides and insulin resistance, independent of insulin concentrations.[107] This suggests that insulin resistance plays a role in the hypertriglyceridemia of type II diabetes mellitus. The removal defect of VLDL in type II diabetes is most likely due to deficiency in lipoprotein lipase. As an increasing number of older patients are accepted for renal transplantation, a greater proportion of the transplant population will have, or develop, type II diabetes. This is likely to increase the incidence of hypertriglyceridemia in renal transplant recipients.

High-Density Lipoprotein

Little data exist on HDL metabolism in transplant recipients. Most investigators have reported normal HDL levels in transplant recipients.[8,13,108–111] The results of two investigations that analyzed HDL_2 and HDL_3 subfractions were conflicting. One study found low HDL_2 and normal HDL_3 levels compared with age-matched controls,[110] while a separate investigator reported high HDL_2 and low HDL_3 levels compared with age- and sex-matched controls.[111]

MANAGEMENT OF HYPERLIPIDEMIA AFTER RENAL TRANSPLANTATION

No prospective study has been done to evaluate the effect of modifying hypercholesterolemia or hypertriglyceridemia on the incidence of vascular disease in renal transplant recipients. In the general population, cholesterol reduction has been shown to be beneficial in reducing the incidence of myocardial infarction in selected populations[18,20–24] and causing regression in established coronary artery lesions.[42,43,112,113] The role of triglyceride reduction has been less well substantiated, partly because hypertriglyceridemia has only recently been accepted as a risk factor for coronary heart disease.[114] However, the triglyceride reduction that occurred in the Helsinki heart study was actually of greater magnitude than the increase in HDL, and some investigators have reinterpreted the data to support the benefit of triglyceride reduction in reducing coronary heart disease events.[115] Given the relatively high incidence of coronary heart disease at the time of transplantation, it is likely that cholesterol reduction would be beneficial in renal transplant recipients with hypercholesterolemia, and possible that triglyceride reduction could also reduce the risk of vascular disease.

Dietary Modification

Dietary modification should be used as the first step in reducing elevated triglyceride and cholesterol levels. Several studies have examined the efficacy of dietary intervention in hyperlipidemic renal transplant recipients (Table 3-1).[96,97,116–118] The studies have included small numbers of patients, very few of whom received cyclosporine. Most of the studies were short-term (i.e., 4 to 13 weeks). Nevertheless, these studies demonstrated that when compliance was good, dietary intervention was effective in reducing triglyceride and cholesterol levels into the normal range. Calorie restriction aimed at reducing body weight was generally successful in reducing serum lipids. Clearly, weight reduction could have major benefits in renal transplant recipients, since most patients become significantly overweight within the first year after transplantation. In one survey of 490 cyclosporine-treated patients, pretransplant relative weight (actual weight compared with ideal body weight) was 102 ± 18 percent (mean ± SD) in females and 100 ± 14 percent in males; at 1 year post-transplant, relative weight had increased to 119 ± 23 percent in females and 110 ± 19 percent in males, and continued to be elevated at 3 years post-transplantation.[17] Diets that have reduced body weight have been uniformly successful in improving the lipid profile in renal transplant recipients (Table 3-1).

Four studies evaluating dietary intervention for post-transplant hypertriglyceridemia showed that triglycerides could be reduced into the normal range with long-term dietary changes.[96,97,116,117] All four studies used a hypocaloric diet that reduced carbohydrate intake. Fat intake was low, with a high polyunsaturated fat to saturated fat ratio (P:S) in three studies[96,97,117]; in one study, fat intake was high.[116] In all studies, weight loss occurred, although it was insignificant in two studies[96,116] and did not reduce relative body weight into the normal range in the third study (relative body weight at the study conclusion was 121 ± 5 percent). In overweight renal transplant recipients, hypertriglyceridemia can be reduced by using a hypocaloric, low-carbohydrate diet regardless of fat content. Dietary efficacy may depend in part on not gaining further weight on the usual hypercaloric post-transplant diet.

The response of hypercholesterolemia to dietary intervention was evaluated in three studies.[96,117,118] All study diets were low cholesterol, hypocaloric, and low fat, with a P : S ratio greater than 1.0. Although fewer than half of the patients complied with the fat reduction, a significant cholesterol-lowering effect was seen. It is possible that with improved dietary compliance, more patients would have been able to achieve normal cholesterol levels.

For patients with combined hypertriglyceridemia and hypercholesterolemia, reduction of carbohydrate and fat intake requires increasing dietary protein. Because a high-protein diet may accelerate the progression of chronic renal failure,[119] an alternative strategy might be to substitute monounsaturated fats for carbohydrate. This diet has been shown to effectively lower cholesterol and triglycerides in healthy men.[120] The composition of fat in the diet remains controversial since some monounsaturated fatty acids raise cholesterol, while some saturated fats do not.[121–123]

Diabetic renal transplant recipients must pay particular attention to serum lipid levels, since they frequently develop both lipid abnormalities and atherosclerotic vascular disease after renal transplantation. Since type II diabetic transplant recipients have insulin resistance exacerbated by corticosteroids and obesity, they may respond to the dietary interventions recommended above. In addition, improved glucose control will decrease hypertriglyceridemia in type II diabetic patients. Type I diabetic patients are more problematic; however, improved glycemic control has been shown to improve hypertriglyceridemia and hypercholesterolemia in non-azotemic diabetic patients[105]; therefore, one goal of dietary intervention should be to normalize glycosylated hemoglobin levels. Dietary interventions to reduce hypertriglyceridemia and hypercholesterolemia in diabetic renal transplant recipients have not been evaluated.

Long-term dietary intervention in renal transplant recipients is limited by a lack of patient compliance. Prior to transplantation, many patients have been anorectic or under multiple dietary restrictions. After transplantation, they are unwilling to continue in such restrictive eating patterns. In a study by Shen et al., 48 percent of patients eligible for dietary intervention declined to participate.[96] Of those who agreed to participate, only 42 percent complied with the recommended dietary fat restrictions. Similar statistics were obtained in other dietary intervention studies.[117,118]

In summary, new renal transplant recipients should be counseled about the likelihood of gaining excessive weight and should be given specific dietary counseling to prevent this occurrence. Overweight transplant recipients with hypertriglyceridemia should be placed on a hypocaloric, low-carbohydrate diet. Overweight renal transplant recipients with hypercholesterolemia should follow a hypocaloric, low-fat, low-cholesterol diet with a P : S ratio greater than 1.0. Whether dietary intervention would be effective in patients immunosuppressed with cyclosporine is not known. Dietary management of post-transplant hyperlipidemia needs to be prospectively evaluated, with particular attention to diabetic patients and patients receiving cyclosporine.

Pharmacologic Management

There are few studies of lipid-lowering agents in renal transplant recipients. Of the currently available agents, most have side effects that may be problematic.[124] Bile acid sequestering agents (cholestyramine, colestipol) are nonabsorbable ion exchange resins that bind bile acids in the intestinal lumen and prevent reabsorption in the terminal ileum. They are unstudied in azathioprine-treated patients and may be contraindicated in cyclosporine-treated patients because of interference with cyclosporine absorption. However, one group of investigators failed to demonstrate an effect of cholestyramine on cyclosporine blood levels in a small number of patients.[125] For treatment of isolated hypercholesterolemia in azathioprine-treated renal transplant recipients, these agents may be useful alone or in combination with other medications. Bile acid sequestrants can increase triglyceride levels and are probably contraindicated in patients with significant hypertriglyceridemia.[126]

Nicotinic acid probably works by decreasing hepatic VLDL production rates, effectively lowering LDL and VLDL levels.[126] Large doses can cause transaminase elevation, hyperuricemia, and glucose intolerance. Since azathioprine may cause hepatotoxicity and cyclosporine may cause hyperuricemia and glucose intolerance,[127,128] nicotinic acid should be used with caution in renal transplant recipients. To date, no studies have examined the safety and efficacy of nicotinic acid after renal transplantation.

Fibric acid derivatives (gemfibrozil and clofibrate) act at multiple sites in the lipoprotein pathway. Clinically, they appear to inhibit VLDL production and enhance VLDL clearance, secondary to stimulation of lipoprotein lipase.[126] In one study, gemfibrozil caused a 20 to 30 percent decrease in plasma triglyceride levels and a concurrent 10 to 20 percent increase in HDL levels.[23] The effect on LDL may be more variable, and gemfibrozil is not recommended as first-line therapy for hypercholesterolemia in the general population.[126] Theoretically, fibric acid analogs would be most appropriate for patients with high serum triglycerides. However, these agents have not been studied in renal transplant recipients. An increased incidence of malignancies in nonazotemic patients,[129] and rhabdomyolysis in patients with nephrotic syndrome,[130] make the use of fibric acid derivatives problematic in renal transplant recipients.

3-Hydroxy-3-methylglutaryl coenzyme A (HMG-CoA) reductase inhibitors are specific competitive inhibitors of the rate-limiting enzyme in cholesterol synthesis. These agents inhibit cholesterol synthesis and thereby induce hepatic LDL receptor synthesis, which leads to enhanced removal of LDL particles.[131] At doses of 20 to 80 mg/d, lovastatin reduced LDL 35 to 45 percent and triglycerides reduced LDL 15 to 20 percent in subjects with familial hypercholesterolemia.[132] Simvastatin and lovastatin effectively lowered cholesterol in nephrotic patients,[133–135] and lovastatin was effective in type II diabetic patients.[136] Therefore, an HMG-CoA reductase inhibitor would be a logical drug for management of renal transplant recipients, who may have hypercholesterolemia secondary to proteinuria and insulin resistance. However, few investigators have evaluated lovastatin in transplant recipients. In 11 renal transplant recipients immunosuppressed with azathioprine, lovastatin reduced LDL cholesterol 28 percent[137]; in 11 cardiac transplant recipients immunosuppressed with cyclosporine, lovastatin reduced LDL cholesterol 34 percent.[138] Potential limitations to lovastatin use in renal transplant recipients include lack of information about long-term safety and reports of rhabdomyolysis in cardiac transplant recipients treated with both lovastatin and cyclosporine.[139–141]

Additional Strategies in the Management of Hyperlipidemia in Renal Transplant Recipients

Dietary modification and lipid-lowering agents constitute the two major approaches to lowering lipid levels in the general population. However, several additional strategies may be beneficial in renal transplant recipients. For example, it may be important to avoid medications that alter lipid levels. Thiazides increase cholesterol and triglyceride levels and β blockers increase triglyceride levels and lower HDL, so alternate antihypertensives should be considered when possible.[141–143] Because of their demonstrated efficacy in decreasing proteinuria,[144] converting enzyme inhibitors might be a good option if antihypertensive drugs are needed. The proteinuria-lowering effect of converting enzyme inhibitors depends on dietary sodium restriction, which the majority of renal transplant recipients are already following. Although multiple studies have demonstrated correlation between the magnitude of proteinuria and serum cholesterol level, there is little data evaluating the response of cholesterol to protein-lowering strategies. Should a converting enzyme inhibitor be introduced, patients should be periodically evaluated to determine whether proteinuria and cholesterol levels have actually decreased with this therapy.

A reduction in the amount of immunosuppressive medication used may have a favorable impact on lipid levels. Stopping prednisone therapy may lower serum cholesterol levels.[75] A change from daily to alternate-day prednisone has been shown to ameliorate hypertriglyceridemia in stable renal transplant recipients in one small series,[7] and to lower both serum cholesterol and triglyceride in a prospective randomized study.[145] Discontinuing cyclosporine therapy 3 months after renal transplantation was associated with a dramatic lowering of cholesterol and triglyceride levels in one study.[13] The risks and benefits of protocols minimizing the amount of immunosuppressive agents must be evaluated in controlled, long-term trials, analyzing both graft survival and incidence of vascular disease.

Alcohol-associated hypertriglyceridemia can be severe. Patients should be advised of this complication, and excessive use of alcohol should be avoided in renal transplant recipients. A regular exercise program has resulted in favorable changes in the lipid profile,[146,147] although this has not been systematically studied in renal transplant recipients. In hypertriglyceridemic men, exercise has been shown to lower triglycerides into the normal range, although no effect on cholesterol levels has been documented. Overall level of fitness has also been associated with a lower risk of death from cardiovascular and coronary heart disease.[148]

Metabolic abnormalities that play a major role in hypertriglyceridemia include diabetes and hyperparathyroidism. As mentioned previously, improved glucose control in type II diabetes has been associated with amelioration of hypertriglyceridemia.[105] Therefore, excellent diabetic control should be achieved when possible. In patients with tertiary hyperparathyroidism, persistent hypertriglyceridemia may be a reason to consider parathyroidectomy in the post-transplant period. In a small series of uremic patients, parathyroidectomy was associated with a dramatic reduction in serum triglyceride level.[104]

SUMMARY

Hyperlipidemia after renal transplantation is a multifactorial problem related to immunosuppressive medications, as well as to the dietary habits and pretransplant lipid status of the patient. Superimposed upon these factors is the possible occurrence of chronic renal failure and proteinuria. Approaches to management include weight reduction, dietary alteration, and attention to the lipid-altering effects of medications required by the patient. Adverse effects from the use of conventional lipid-lowering agents are likely to be especially frequent in renal transplant recipients. Although newer antilipemic agents may be more effective in treating post-transplant hyperlipidemia, the determination of their safety and efficacy will require careful study.

REFERENCES

1. Ghosh P, Evans DB, Tomlinson SA, Calne RY: Plasma lipids following renal transplantation. Transplantation 15:522, 1974
2. Casaretto A, Marchioro TL, Goldsmith R, Bagdade JD: Hyperlipidemia after successful renal transplantation. Lancet 1:481, 1974
3. Bagdade JD, Casaretto A, Albers JJ: Effects of chronic uremia, hemodialysis and renal transplantation on plasma lipids and lipoproteins in man. J Lab Clin Med 87:37, 1976
4. Salanda LF, Hurst KS, Amend WJC et al: Hyperlipidemia after renal transplantation in children. Am J Dis Child 130:951, 1976
5. Ibels LS, Alfrey AC, Weil R: Hyperlipidemia in adult, pediatric and diabetic renal transplant recipients. Am J Med 64:634, 1978
6. Ponticelli C, Barbi GL, Cantaluppi A et al: Lipid disorders in renal transplant recipients. Nephron 20:189, 1978
7. Cattran DC, Steiner G, Wilson DR, Fenton SSA: Hyperlipidemia after renal transplantation: natural history and pathophysiology. Intern Med 91:554, 1979

8. Gokal R, Mann JI, Moore RA, Morris PJ: Hyperlipidemia following renal transplantation. Q J Med 64:634, 1979
9. Nicholls AJ, Cumming AM, Catto GRD et al: Lipid relationships in dialysis and renal transplant patients. Q J Med 198:149, 1981
10. Chan MK, Varghese Z, Moorehead JF: Lipid abnormalities in uremia, dialysis, and transplantation. Kidney Int 19:625, 1981
11. Chan MK, Varghese Z, Persauel JW et al: The role of multiple pharmacotherapy in the pathogenesis of hyperlipidemia after renal transplantation. Clin Nephrol 15:309, 1981
12. Harris KPJ, Russell GI, Parvin SD et al: Alterations in lipid and carbohydrate metabolism attributable to cyclosporin A in renal transplant recipients. Br Med J 292:16, 1986
13. Lowry RP, Soltys G, Mangel R et al: Type II hyperlipoproteinemia, hyperapobetalipoproteinemia, and hyperalphalipoproteinemia following renal transplantation: prevalence and precipitating factors. Transplant Proc 19:2229, 1987
14. Kasiske BL, Umen AJ: Persistent hyperlipidemia in renal transplant patients. Medicine 66:309, 1987
15. Raine AEG, Carter R, Mann JI, Morris PJ: Adverse effect of cyclosporin on plasma cholesterol in renal transplant recipients. Nephrol Dial Transplant 3:458, 1988
16. Markell MS, Brown CD, Butt KMH, Friedman EA: Prospective evaluation of changes in lipid profiles in cyclosporin-treated renal transplant patients. Transplant Proc 21:1497, 1989
17. Vathsala A, Weinberg RB, Schoenberg L et al: Lipid abnormalities in cyclosporin-prednisone-treated renal transplant recipients. Transplantation 48:37, 1989
18. Leren P: The Oslo Diet-Heart Study: eleven year report. Circulation 42:935, 1970
19. Kannel WB, Castelli WP, Gordon T, McNamara PM: Serum cholesterol, lipoproteins and the risk of coronary heart disease: the Framingham study. Ann Intern Med 74:1, 1971
20. Coronary Drug Project Research Group: Clofibrate and niacin in coronary heart disease. JAMA 231:360, 1975
21. Lipid Research Clinics Program: The Lipid Research Clinics Coronary Primary Prevention Trial results. I. Reduction in incidence of coronary heart disease. JAMA 251:351, 1984
22. Tyroler HA: Review of lipid-lowering clinical trials in relation to observational epidemiologic studies. Circulation 76:515, 1987
23. Frick MH, Elo O, Haapa K et al: Helsinki Heart Study: Primary prevention trial in middle-aged men with dyslipidemia. N Engl J Med 317:1237, 1987
24. Pekkanen J, Linn S, Heiss G et al: Ten-year mortality from cardiovascular disease in relation to cholesterol level among men with and without preexisting cardiovascular disease. N Engl J Med 322:1700, 1990
25. Ibels LS, Stewart JH, Mahony JF, Sheil AGR: Deaths from occlusive arterial disease in renal allograft recipients. Br Med J 3:552, 1974
26. Kirkman RL, Strom TB, Weir MR et al: Late mortality and morbidity in recipients of long-term renal allografts. Transplantation 34:347, 1982
27. Washer GF, Schröter GPJ, Starzl TE, Weil R: Causes of death after kidney transplant. JAMA 250:49, 1983
28. Mahony JF, Savdie E, Caterson RJ et al: The natural history of renal allografts beyond ten years. Transplant Proc 18:135, 1986

29. Fassbinder W, Challah S, Brynger H (EDTA Registry): Long-term results of renal transplantation in Europe. Transplant Proc 19:3754, 1987
30. Fabrega A, Matas AJ, Payne WD et al: Ten to twenty year followup of 123 consecutive HLA-identical living related kidney transplants from the pre-cyclosporine era. Clin Transplantation 4:145, 1990
31. Grundy SM, Vega GL: Cause of high blood cholesterol. Circulation 81:412, 1990
32. Gordon DJ, Rifkind BM: High-density lipoprotein: the clinical implications of recent studies. N Engl J Med 321:1311, 1989
33. Fielding CJ: Factors affecting the rate of catalyzed transfer of cholesterol esters in plasma. Am Heart J 113:532, 1987
34. Brown MS, Goldstein JL: A receptor-mediated pathway for cholesterol homeostasis. Science 232:34, 1986
35. Eisenberg S: High density lipoprotein metabolism. J Lipid Res 25:1017, 1984
36. Nikkila EA, Taskinen M-R, Kekki M: Relation of plasma high-density lipoprotein cholesterol to lipoprotein-lipase activity in adipose tissue and skeletal muscle of man. Atherosclerosis 29:497, 1978
37. Eckel RH: Lipoprotein lipase. N Engl J Med 320:1060, 1989
38. Fielding PE, Fielding CJ, Havel RJ et al: Cholesterol net transport, esterification, and transfer in human hyperlipidemic plasma. J Clin Invest 71:449, 1983
39. Steinberg D: Lipoproteins and the pathogenesis of atherosclerosis. Circulation 76:508, 1987
40. Henrikson T, Mahoney EM, Steinberg D: Enhanced macrophase degradation of low density lipoprotein previously incubated with cultured endothelial cells. Proc Natl Acad Sci USA 78:6499, 1981
41. Carvalho CA, Colman RW, Lees RS: Platelet function in hyperlipoproteinemia. N Engl J Med 29:434, 1974
42. Nikkilä EA, Viikinkoski P, Valk M et al: Prevention of progression of coronary atherosclerosis by treatment of hyperlipidemia: a seven year prospective angiographic study. Br Med J 289:220, 1984
43. Blankenhorn DH, Nessum SA, Johnson RL et al: Beneficial effects of combined colestipol-niacin therapy on coronary atherosclerosis and coronary bypass grafts. JAMA 257:3233, 1987
44. Weinrauch L, D'Elia JA, Healy RW et al: Asymptomatic coronary artery disease: angiographic assessment of diabetics evaluated for renal transplantation. Circulation 58:1184, 1978
45. Braun WE, Philips DF, Vidt DG et al: Coronary artery disease in 100 diabetics with end-stage renal failure. Transplant Proc 16:603, 1984
46. Philipson JD, Carpenter BJ, Itzkoff J et al: Evaluation of cardiovascular risk for renal transplantation in diabetics. Am J Med 81:630, 1986
47. Lorber MI, VanBuren CT, Flechner SM et al: Pretransplant coronary arteriography for diabetic renal transplant recipients. Transplant Proc 19:1539, 1987
48. Gunnarsson R, Löfmark R, Norlander R et al: Acute myocardial infarction in renal transplant recipients: incidence and prognosis. Eur Heart J 5:218, 1984
49. Jensen T, Borch-Johnsen K, Kofoed-Enevoldsen A et al: Coronary heart disease in young type I (insulin-dependent) diabetic patients with and without diabetic nephropathy: incidence and risk factors. Diabetologia 30:144, 1987
50. Vincenti F, Amend WJ, Abele J et al: The role of hypertension in hemodialysis-associated athersclerosis. Am J Med 68:366, 1980

51. Rostrand SG, Kirk KA, Rutsky EA: Relationship of coronary risk factors to hemodialysis-associated ischemic heart disease. Kidney Int 22:304, 1982
52. Ritz E, Strumpf C, Katz F et al: Hypertension and cardiovascular risk factors in hemodialyzed diabetic patients. Hypertension suppl. II 7 II-118, 1985
53. Parving HH, Anderson AR, Smidt UM et al: Diabetic nephropathy and arterial hypertension. Diabetologia 24:10, 1983
54. Bagdade JD, Albers JJ: Plasma high-density lipoprotein concentration in chronic hemodialysis and renal transplant patients. N Engl J Med 296:1436, 1977
55. Frank WM, Rao TKS, Manis T et al: Relationship of plasma lipids to renal function and length of time on maintenance hemodialysis. Am J Clin Nutr 31:1886, 1978
56. Goldberg AP, Applebaum-Bowden DM, Bierman EL et al: Increase in lipoprotein lipase during clofibrate treatment of hypertriglyceridemia in patients on hemodialysis. N Engl J Med 301:1073, 1979
57. Nestel PJ, Fidge NH, Tan MH: Increased lipoprotein-remnant formation in chronic renal failure. N Engl J Med 307:329, 1982
58. Chan MK, Persaud J, Varghese Z et al: Pathogenic roles of post-heparin lipases in lipid abnormalities in hemodialysis patients. Kidney Int 25:812, 1984
59. Hahn R, Oette K, Mondorf H et al: Analysis of cardiovascular risk factors in chronic hemodialysis patients with special attention to the hyperlipoproteinemias. Atherosclerosis 48:279, 1983
60. Goldberg AP, Harter HR, Patsch W et al: Racial differences in plasma high-density lipoproteins in patients receiving hemodialysis. N Engl J Med 308:1245, 1983
61. Altman P, Alaupovic P, Gustafson A: Serum apolipoproteins of patients with chronic renal failure. Kidney Int 32:368, 1987
62. Fuh MMT, Lee CM, Jeng CY et al: Effect of chronic renal failure on high-density lipoprotein kinetics. Kidney Int 37:1295, 1990
63. Appel GB, Blum CB, Chien S et al: The hyperlipidemia of nephrotic syndrome. N Engl J Med 312:1544, 1985
64. Sokolovskaya IV, Nikiforova NV: High-density lipoprotein cholesterol in patients with untreated and treated nephrotic syndrome. Nephron 37:49, 1984
65. Kaysen GA, Gambertoglio J, Felts J et al: Albumin synthesis, albuminuria and hyperlipidemia in nephrotic patients. Kidney Int 31:1368, 1987
66. Joven J, Villabona C, Vilella E et al: Abnormalities of lipoprotein metabolism in patients with the nephrotic syndrome. N Engl J Med 323:579, 1990
67. Zavaroni I, Bonora E, Pagliara M: Risk factors for coronary artery disease in healthy persons with hyperinsulinemia and glucose intolerance. N Engl J Med 320:702, 1989
68. Foster DW: Insulin resistance—a secret killer? N Engl J Med 320:733, 1989
69. Cerletty JM, Engbring NH: Azotemia and glucose intolerance. Ann Intern Med 66:1097, 1967
70. Jaspan JB, Mako ME, Kuzoya H et al: Abnormalities in circulating beta cell peptides in chronic renal failure: comparison of C-peptide, proinsulin and insulin. J Clin Endocrinol Metab 45:441, 1977
71. Bagdade JD, Subbaiah PV, Bartos D et al: Polyamines: an unrecognized cardiovascular risk factor in chronic dialysis? Lancet 1:412, 1979
72. Krolewski AS, Kosinski EJ, Warram JH et al: Magnitude and determinants of coronary artery disease in juvenile-onset, insulin-dependent diabetes mellitus. Am J Cardiol 59:750, 1987

73. Thomas FT, Lee HM: Factors in the differential rate of arteriosclerosis between long-surviving renal transplant recipients and dialysis patients. Ann Surg 184:342, 1976
74. Kasiske BL: Risk factors for accelerated atherosclerosis in renal transplant recipients. Am J Med 84:985, 1988
75. Reisman L, Lieberman KV, Burrows L, Schanzer H: Follow-up of cyclosporine-treated pediatric renal allograft recipients after cessation of prednisone. Transplantation 49:76, 1990
76. Becker DM, Chamberlain B, Swank R et al: Relationship between corticosteroid exposure and plasma lipid levels in heart transplant recipients. Am J Med 85:632, 1988
77. Taskinen M-R, Nikkila EA, Pelkonen R, Sane T: Plasma lipoproteins, lipolytic enzymes, and very-low-density-lipoprotein triglyceride turnover in Cushing's syndrome. J Clin Endocrinol Metab 57:619, 1983
78. Stern MP, Kolterman JF, Fries JF et al: Adrenocortical steroid treatment of rheumatic diseases. Arch Intern Med 132:97, 1973
79. Bagdade JD, Yee E, Albers J, Pykalisto OS: Glucocorticoids and triglyceride transport: effects on triglyceride secretion rates, lipoprotein lipase and plasma lipoproteins in the rat. Metabolism 25:533, 1976
80. Weinberg R: Lipoprotein metabolism: hormonal regulation. Hosp Pract 22:125, 1987
81. Hirsch LJ, Mazzone T: Dexamethasone modulates lipoprotein metabolism in cultured human monocyte-derived macrophages. J Clin Invest 77:485, 1986
82. Ballantyne CM, Podet EJ, Patsch WP et al: Effect of cyclosporin therapy on plasma lipoprotein levels. JAMA 262:53, 1989
83. Ellis CN, Gorsulowsky DC, Hamilton TA et al: Cyclosporin improves psoriasis in a double-blind study. JAMA 256:3110, 1986
84. Lemaire M, Tillement P: Role of lipoproteins and erythrocytes in the in vitro binding and distribution of cyclosporin A in the blood. J Pharm Pharmacol 34:715, 1982
85. DeGroen P: Cyclosporin, low-density lipoprotein, and cholesterol. Mayo Clin Proc 63:1012, 1988
86. Markell MS, Friedman A: Hyperlipidemia after organ transplantation. Am J Med 87:62N, 1989
87. Princen HMG, Meijer P, Hofstee B et al: Effects of cyclosporin A on LDL receptor activity and bile acid synthesis in hepatocyte monolayer cultures and in vivo in the rat, abstracted. Hepatology 7:1109, 1987
88. Hyperlipidemia after renal transplantation (editorial). Lancet 1:919, 1988
89. Jackson JM, Lee HA: The role of propranolol therapy and proteinuria in the etiology of post renal transplantation hyperlipidemia. Clin Nephrol 18:95, 1982
90. Manske CL: Lipid abnormalities and renal disease. Kidney 20:25, 1988
91. Marsh JB, Sparks CE: Hepatic secretion of lipoproteins in the rat and the effects of experimental nephrosis. J Clin Invest 64:1229, 1979
92. Vega GL, Grundy SM: Lovastatin therapy in nephrotic hyperlipidemia: effects on lipoprotein metabolism. Kidney Int 33:1160, 1988
93. Markell MS, Brown CD, Friedman EA: Risk of early post-transplant hyperlipidemia is related to pre-transplant LDL level, abstracted. Kidney Int 35:519, 1989
94. Ost L, Tyden G, Fehrman I: Impaired glucose tolerance in cyclosporin treated renal graft recipients. Transplantation 45:1027, 1988

95. Jensen SL, Falholt K, Hagen K, Nielson OV: Effects of cyclosporine A and prednisone on glucose and lipid metabolism. Transplant Proc 19:4011, 1987
96. Shen SY, Lukens CW, Alongi SV et al: Patient profile and effect of dietary therapy on post-transplant hyperlipidemia. Kidney Int, suppl. 24: S-147, 1983
97. Nelson J, Beauregard H, Gélinas M: Rapid improvement of hyperlipidemia in kidney transplant patients with a multifactorial hypolipidemic diet. Transplant Proc 20:1264, 1988
98. Hahn R, Oette K, Mondorf H et al: Analysis of cardiovascular risk factors in hemodialysis patients with special attention to the hyperlipoproteinemias. Atherosclerosis 48:279, 1983
99. Frank WM, Rao TKS, Manis T et al: Relationship of plasma lipids to renal function and length of time on dialysis. Am J Clin Nutr 31:1886, 1978
100. Nestel PJ, Fidge NH, Tan MH: Increased lipoprotein-remnant formation in chronic renal failure. N Engl J Med 307:329, 1982
101. Chan MK, Persuad J, Varghese Z et al: Pathogenic roles of post-heparin lipases in lipid abnormalities in hemodialysis patients. Kidney Int 25:812, 1984
102. Goldberg AP, Appelbaum-Bowden DM, Bierman EL et al: Increase in lipoprotein lipase during clofibrate treatment of hypertriglyceridemia in patients on hemodialysis. N Engl J Med 301:1073, 1979
103. Akmal M, Kasim SE, Soliman AR, Massry SG: Excess parathyroid hormone adversely affects lipid metabolism in chronic renal failure. Kidney Int 37:854, 1990
104. Lacour B, Roullet J-B, Liagre A-M et al: Serum lipoprotein disturbances in primary and secondary hyperparathyroidism and effects of parathyroidectomy. Am J Kidney Dis 8:422, 1986
105. Howard BV: Lipoprotein metabolism in diabetes mellitus. J Lipid Res 28:613, 1987
106. Dunn FL, Carroll P, Vlachokostaf F et al: Effect of treatment with the artificial beta-cell on triglyceride metabolism in type I diabetes mellitus. Diabetes suppl. 1 34:86A, 1985
107. Abbott WGH, Lillioja AA, Young JK et al: Relationship between plasma lipoprotein concentration and insulin action. Diabetes 38:897, 1987
108. Curtis JJ, Galla JH, Woodford SY et al: Effects of renal transplantation on hyperlipidemia and high-density lipoprotein cholesterol. Transplantation 26:364, 1978
109. Jung K, Neumann R, Scholz D, Nugel E: Abnormalities in the composition of serum high density lipoprotein in renal transplant recipients. Clin Nephrol 17:191, 1982
110. Ettinger WH, Bender WL, Goldberg AP, Hazzard WR: Lipoprotein abnormalities in healthy renal transplant recipients: persistence of low HDL_2 cholesterol. Nephron 47:17, 1987
111. Jung K, Scheifler A, Blank W et al: Changed composition of high-density lipoprotein subclasses HDL_2 and HDL_3 after renal transplantation. Transplantation 46:409, 1988
112. Levy RI, Brensike JF, Epstein SE et al: The influence of changes in lipid values induced by cholestyramine and dict on progression of coronary artery disease: results of NHLBI Type II Coronary Intervention Study. Circulation 69:325, 1984
113. Arntzenius AC, Kromhout D, Barth JD et al: Diet, lipoproteins, and progression of coronary atherosclerosis: the Leiden Intervention Trial. N Engl J Med 312:805, 1985

114. Blankenhorn DH, Alaupovic P, Wickham E et al: Prediction of angiographic change in native human coronary arteries and aortocoronary bypass grafts: lipid and nonlipid factors. Circulation 81:470, 1990
115. Havel R: Role of triglyceride-rich lipoproteins in progression of atherosclerosis (editorial). Circulation 81:94, 1990
116. Ponticelli C, Cantaluppi A, Barbi G et al: Treatment of hypertriglyceridemia after renal transplantation. Br Med J 2:597, 1975
117. Disler PB, Goldberg RB, Kuhn L et al: The role of diet in the pathogenesis and control of hyperlipidemia after renal transplantation. Clin Nephrol 16:29, 1981
118. Moore RA, Callahan MF, Cody M: The effect of the American Heart Association Step One Diet on hyperlipidemia following renal transplantation. Transplantation 49:60, 1990
119. Brenner BM, Meyer TW, Hostetter TH: Dietary protein intake and the progressive nature of kidney disease. N Engl J Med 307:652, 1982
120. Ginsberg HN, Barr SL, Gilbert A et al: Reduction of plasma cholesterol levels in normal men on an American Heart Association step 1 diet or a step 1 diet with added monounsaturated fat. N Engl J Med 322:574, 1990
121. Bonanome A, Grundy SM: Effect of dietary stearic acid on plasma cholesterol and lipoprotein levels. N Engl J Med 318:1244, 1988
122. Mensink RP, Katan MB: Effect of dietary trans fatty acids on high-density and low-density lipoprotein cholesterol levels in healthy subjects. N Engl J Med 323:439, 1990
123. Grundy SM: Trans monounsaturated fatty acids and serum cholesterol levels (editorial). N Engl J Med 323:480, 1990
124. Grundy SM: Management of hyperlipidemia of kidney disease. Kidney Int 37:847, 1990
125. Keogh A, Day R, Critchley L et al: The effect of food and cholestyramine on the absorption of cyclosporine in cardiac transplant recipients. Transplant Proc 20:27, 1988
126. Witztum JL: Current approaches to drug therapy for the hypercholesterolemic patients. Circulation 80:1101, 1988
127. Lin H-Y, Rocher LL, McQuillan MA: Cyclosporin-induced hyperuricemia and gout. N Engl J Med 32:287, 1989
128. Kahan BD: Cyclosporine. N Engl J Med 321:1725, 1989
129. WHO Cooperative Trial on Primary Prevention of Ischaemic Heart Disease Using Clofibrate to Lower Serum Cholesterol: Mortality follow-up report. Lancet 2:379, 1980
130. Bridgeman JF, Rosen SM, Thorp JM: Complications during clofibrate treatment of nephrotic syndrome hyperlipoproteinemia. Lancet 2:506, 1972
131. Grundy SM: Drug therapy: HMG-CoA reductase inhibitors for treatment of hypercholesterolemia. N Engl J Med 319:24, 1988
132. Havel RJ, Hunninghake DB, Illingworth DR et al: Lovastatin in the treatment of heterozygous familial hypercholesterolemia: a multicenter study. Ann Intern Med 107:609, 1987
133. Rabelink AJ, Hené RJ, Erkelens DW et al: Effects of simvastatin and cholestyramine on lipoprotein profile in hyperlipidemia of nephrotic syndrome. Lancet 2:1335, 1988
134. Vega GL, Grundy JM: Lovastatin therapy in nephrotic hyperlipidemia: effects on lipoprotein metabolism. Kidney Int 33:1160, 1988

135. Kasiske BL, Velosa JA, Halstenson CE et al: The effects of lovastatin in hyperlipidemic patients with nephrotic syndrome. Am J Kidney Dis 15:8, 1990
136. Garg A, Grundy SM: Lovastatin for lowering cholesterol levels in non-insulin-dependent diabetes mellitus. N Engl J Med 318:81, 1988
137. Kasiske BL, Tortorice KL, Heim-Duthoy KL et al: Lovastatin treatment of hypercholesterolemia in renal transplant recipients. Transplantation 49:95, 1990
138. Kuo PC, Kirshenbaum JM, Gorden J et al: Lovastatin therapy for hypercholesterolemia in cardiac transplant recipients. Am J Cardiol 64:631, 1989
139. Corpier CL, Jones PH, Suki WN et al: Rhabdomyolysis and renal injury with lovastatin use. JAMA 260:239, 1988
140. Norman DJ, Illingworth DR, Munson J, Hosenpud J: Myolysis and acute renal failure in a heart-transplant recipient receiving lovastatin. N Engl J Med 318:46, 1988
141. East C, Alivazatos PA, Grundy SM et al. Rhabdomyolysis in patients receiving lovastatin after cardiac transplantation. N Engl J Med 318:47, 1988
142. Roberts WC: Recent studies on the effects of beta blockers on blood lipid levels. Am Heart J 117:709, 1989
143. Recommendations of the Adult Treatment Panel of the National Cholesterol Education Program: Implications for the management of hypertension (editorial). Hypertension 12:471, 1988
144. Heeg JE, DeJong EP, Van Der Hem GK, De Zeeuw D: Efficacy and variability of the antiproteinuric effects of ACE inhibition by lisinopril. Kidney Int 36:272, 1989
145. Curtis JJ, Galla JH, Woodford SY et al: Effect of alternate-day prednisone on plasma lipids in renal transplant recipients. Kidney Int 22:42, 1982
146. Superko HR, Haskell WH: The role of exercise training in the therapy of hyperlipoproteinemia. Cardiol Clin 5:285, 1987
147. Wood PD, Stefanick ML, Dreon DM et al: Changes in plasma lipids and lipoproteins in overweight men during weight loss through dieting as compared with exercise. N Engl J Med 319:1173, 1988
148. Ekelund L-G, Haskell WL, Johnson JL et al: Physical fitness as a predictor of cardiovascular mortality in asymptomatic North American men. N Engl J Med 319:1379, 1988
149. Havel RJ: Approach to the patient with hyperlipidemia. Med Clin North Am 66:2, 1982

4

Lipid Metabolism in the Nephrotic Syndrome

Jaap A. Joles
George A. Kaysen

INTRODUCTION

The nephrotic syndrome, characterized by proteinuria in excess of 3.5 g/d, hypoproteinemia, edema formation, and hyperlipidemia, is a consequence of altered permselectivity of the glomerular filtration barrier and may occur in the course of a variety of systemic diseases or diseases restricted to the kidney alone. It is not intuitively obvious why an alteration in the glomerular filtration barrier and the resulting urinary protein loss should profoundly affect lipid metabolism.

There is general agreement that in experimental models of the nephrotic syndrome, hepatic synthesis of both lipids and apolipoproteins[1] are increased, while removal of lipoproteins, chylomicrons (CMs),[2] and very low-density lipoproteins (VLDLs)[3] is reduced. The precise cause and relative contribution of increased lipogenesis and decreased lipid catabolism to hyperlipidemia, and their relationship to urinary protein loss, hypoalbuminemia, and reduced plasma oncotic pressure (π) remains controversial. The defect in the metabolism of plasma lipoproteins in the nephrotic syndrome resulting in increased plasma triglycerides (TGs) and cholesterol affects virtually every class of lipoproteins, not only altering their plasma concentration, but also their composition and metabolic rate. Two separate mechanisms have been proposed to establish a relationship between proteinuria and altered lipid metabolism. One postulates that reduced plasma π, caused by proteinuria, directly stimulates the synthesis of lipids and lipoproteins by the liver.[4] This model is directed at explaining the mechanism of increased lipogenesis and is not focused on the mechanisms of deranged lipid removal, although it has been proposed that increased production of VLDL may saturate the mechanisms responsible for catabolism of TG-rich lipoproteins and thus cause reduced clearance of these particles directly.[5] A second proposed mechanism is that TG catabolism is impeded as a result of reduced plasma albumin concentration[6] or as a result of the urinary loss of a liporegulatory substance.[2] The latter mechanisms do not directly address the question of increased lipogenesis.

SYNTHESIS AND CATABOLISM OF CIRCULATING LIPOPROTEINS

VLDL is synthesized in the liver (Fig. 4-1) and CMs are synthesized in the gut (Fig. 4-2); however, both particles are quite similar, transporting fatty acids to muscles, heart, adipose tissue, and to liver, and will be considered together. Both of these large TG-laden lipoproteins are acted on by lipoprotein lipase (LPL) at the capillary endothelium (Figs. 4-1 and 4-2), and not at the site of fatty acid absorption. Liberated free fatty acids must be transported from the site of hydrolysis to the site of absorption at the fat cell or muscle cell. Since albumin binds free fatty acids and reduces their local concentration at the site of hydrolysis,[6] one mechanism whereby hypoalbuminemia may retard lipid catabolism is by allowing an increase in free fatty acid concentration to occur locally, resulting in inhibition of LPL[7]. VLDL is ultimately catabolized either to low-density lipoprotein (LDL) or to intermediate-density lipoprotein (IDL) and CMs to remnant particles, all of which are finally removed by the liver.

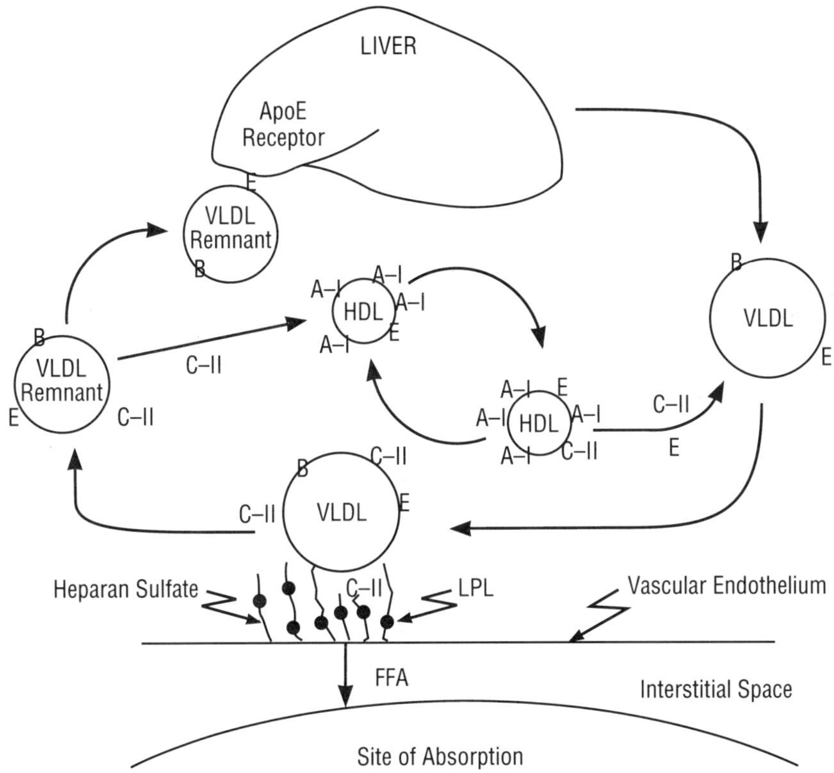

Fig. 4-1. Representation of catabolic pathways affecting VLDL. Lipoprotein lipase (LPL) is tethered to the vascular endothelium by heparan sulfate and achieves full activity only in the presence of apolipoprotein C-II. Liberated free fatty acid (FFA) must traverse the interstitial space to the site of absorption by muscle or adipose cells. A-I, B, C-II, and E represent apolipoproteins A-I, B, C-II, and E, respectively. HDL, high-density lipoprotein.

LPL requires apolipoprotein C-II (apo C-II) for full activity.[8] This apolipoprotein is presented to LPL on the surface of VLDL or CMs at the time of hydrolysis (Figs. 4-1 and 4-2). High-density lipoprotein (HDL) serves the function of shuttling apo C-II from remnants to VLDL and CMs in order to allow their hydrolysis to proceed. Normal catabolism of both VLDL and CMs therefore requires the presence of normally functioning HDL, mole for mole the most abundant of plasma lipoproteins.

HDL is derived from apolipoproteins synthesized either in the liver or gut, cholesterol, cholesterol esters, and phospholipids. This particle initially appears as a discoid nascent HDL particle containing little or no cholesterol esters.[9] Cholesterol is esterified by the enzyme lecithin cholesterol acyltransferase (LCAT).[9,10] In this reaction, phospholipids are hydrolyzed. The fatty acid moiety, usually arachadonic acid, is combined with cholesterol to form the cholesterol ester and a mole of lysolecithin is liberated. As in the

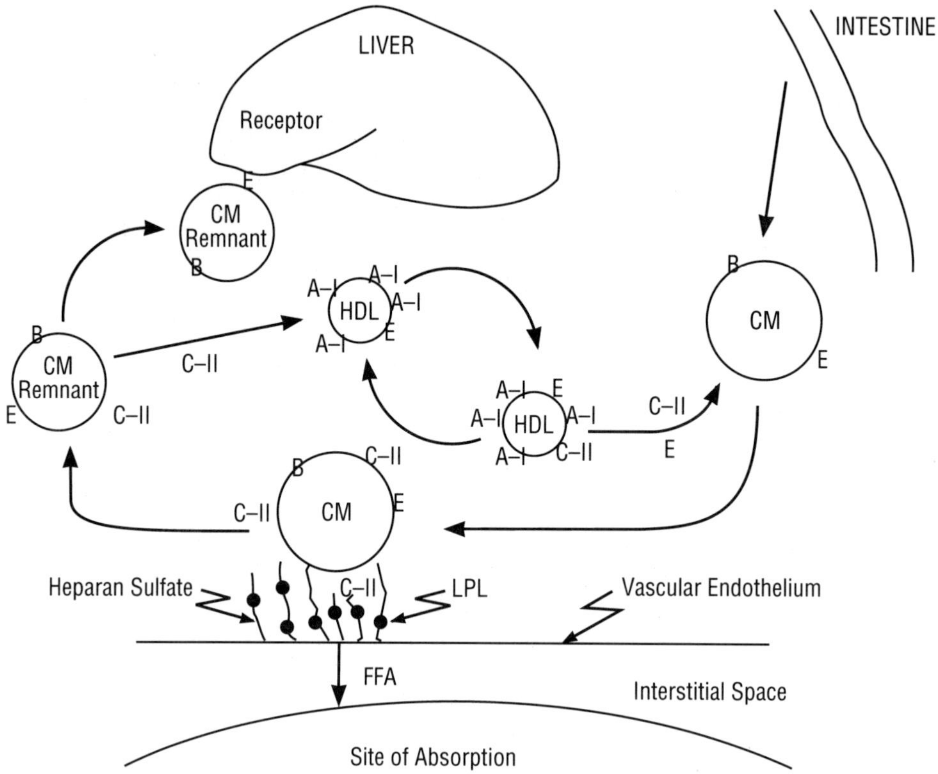

Fig. 4-2. Representation of catabolic pathways affecting CMs. Abbreviations are as in Figure 4-1.

case of fatty acid transport, albumin serves to bind liberated lysolecithin and accelerates activity of LCAT, thus facilitating the maturation of HDL.[11] The more hydrophobic cholesterol esters sink into the core of the nascent HDL particle, changing its conformation to spherical HDL_3 and ultimately to larger, less dense HDL_2. Mature HDL_2 is necessary for transport of the essential LPL cofactor apo C-II to VLDL and to CMs. Apo C-II is recycled by HDL_2 from both CMs and VLDL as they are catabolized by LPL (Figs. 4-1 and 4-2). Without recycling of apo C-II by HDL_2, the action of LPL on these large lipoproteins will be greatly reduced. The alteration in HDL composition observed in the nephrotic syndrome is precisely that anticipated to occur should LCAT activity be reduced.

The cholesterol moiety of HDL originates in part in CMs and VLDL and is transferred to HDL[12] when these particles are acted up by LPL. Decreases in LPL activity can therefore cause abnormalities in HDL composition as well as be caused by abnormalities in the composition of this lipoprotein.

PLASMA LIPIDS AND LIPOPROTEIN COMPOSITION IN THE NEPHROTIC SYNDROME

Hyperlipidemia is found in nearly all nephrotic patients with an albumin concentration below 20 g/L.[13,14] Cholesterol and phospholipids correlate closely with one another, but cholesterol and TGs do not,[14] indicating that multiple disturbances in lipid metabolism may exist in nephrotic patients. Although changes in plasma concentrations of apo A-I and apo A-II are variable,[15,16] apo B is increased consistently.[15,16]

The characteristic disorder in blood lipoprotein composition in nephrotic patients is an increase in the LDL, VLDL, and/or IDL fractions, and a decrease in HDL resulting in an increase in the LDL to HDL cholesterol ratio.[16–18] HDL subtypes found in plasma of nephrotic patients are also abnormally distributed[17]: HDL_3 is modestly elevated, while HDL_2 is markedly reduced. Since it is primarily the latter subclass of HDL that is protective against atherosclerosis, the combination of reduced HDL_2 in conjunction with increased LDL, VLDL, and IDL cholesterol potentially poses significant risk for cardiovascular disease.

In addition to these quantitative defects in lipoprotein distribution, a qualitative defect also occurs within the lipoprotein fractions of nephrotic patients.[16] The proportion of cholesterol esters to free cholesterol, of phospholipids to TGs, and of lipids to protein is increased in each lipoprotein fraction compared with non-nephrotic control patients.[16] Lipoprotein particles, rich in phospholipids and esterified cholesterol and resembling of remnants of both VLDL and CM accumulate, possibly as the result of a reduced ability to convert VLDL to IDL or LDL, or to reduced uptake of these latter particles by the liver.

The nephrotic syndrome may also increase the risk of atherosclerotic disease by altering the level of Lp(a).[19] Lp(a) is a lipoprotein that consists of a glycoprotein apo(a) synthesized in the liver that circulates bound to apo B_{100} on LDL.[20] Lp(a) may be a powerful independent risk factor for atherosclerotic disease in humans.[21,22] Lp(a) levels are determined genetically and do not vary with diet or with plasma cholesterol levels,[23] although levels of this protein may be increased as part of the acute phase response[24] or in association with hypertriglyceridemia.[20] Lp(a) has been shown to be increased significantly in a group of nephrotic patients compared with non-nephrotic control subjects,[19] and was elevated to clinically significant values ($>$ 300 mg/L) in a few patients with proteinuria due to membranous glomerulonephritis (n=3) and primary amyloidosis (n=3).

Lp(a) is not found in common laboratory animals used to study the nephrotic syndrome. Despite this and several other differences in lipid metabolism in humans and rats,[25] the nephrotic syndrome in both species affects lipid metabolism similarly. Several models of the nephrotic syndrome in the rat have therefore provided important insights into the pathophysiology of this disease process in humans.

One of the most notable differences between the two species is the presence of very large amounts of HDL in the rat and only very small amounts of LDL. Furthermore, only very low levels of HDL_3 are found in the rat, because of an active LCAT system rapidly converting HDL_3 to HDL_2 and ultimately to an even larger apo E-rich particle, HDL_1, a lipoprotein found in only small amounts in humans. Most cholesterol is transported in HDL in the rat as opposed to LDL in humans, and total cholesterol levels are very low in normal rats (100 mg/dl). In addition, VLDL is cleared very rapidly from the plasma of normal rats ($t\frac{1}{2}$ 8 to 10 minutes)[26] when compared with humans ($t\frac{1}{2}$ of several hours).[27]

The proportion of TG to total protein is greatly increased both in CMs and in VLDL particles in the rat with nephrotic syndrome induced by puromycin aminonucleoside (PAN).[5] VLDL derived from livers isolated from nephrotic rats show similar defects in compositon,[28] suggesting that these alterations may occur in part prior to processing of VLDL in plasma. Chylomicrons and VLDL have a larger core and contain less apolipoprotein per particle.[5] This decrease is due to reduced incorporation of apo A-I, apo B, apo C, and apo E. In addition, apolipoproteins are incorporated into CMs at a lower rate than in controls[29] when labeled HDL is injected into nephrotic rats. Thus, a defect either in the composition of CMs or in the interaction between HDL and other lipoproteins occurs in the nephrotic syndrome.[28] The amount of apo E, a protein recognized by a specific receptor on the hepatocyte necessary for the uptake of IDL and CM remnants, and apo A-I are both greatly reduced per molecule of CM and VLDL.[5] Apo C-II is also reduced in concentration per molecule of VLDL.[30] This reduction in apo C-II may play a role in the reduced rate of catabolism of VLDL, although decreased activity of LPL may not be the principal factor responsible for the reduced clearance of either VLDL or CMs.[5,26]

Plasma HDL in nephrotic rats contains mostly apo A-I (90 percent of total), but little apo C, apo A-IV, or apo E in comparison to normal rats.[31,32] As in the case of VLDL, HDL derived from perfused livers isolated from nephrotic rats shows qualitative abnormalities.[28] These qualitative differences in the composition of CMs, VLDL, and HDL greatly influence catabolism of these lipoproteins and thus have important pathophysiologic consequences that will be detailed below.

INCREASED LIPID SYNTHESIS IN THE NEPHROTIC SYNDROME

It has been well established that hepatic synthesis[33,34] and secretion of lipoproteins[1,4,28,35] are increased in nephrotic syndrome in the rat.

The rate of secretion of VLDL is increased twofold in perfused livers from rats with nephrotic syndrome induced by the administration of PAN,[28,35] but only 20 percent in vivo in nephrotic rats fasted for 5 hours.[36] Fasting for 24 hours reduces VLDL secretion by approximately 70 percent in nephrotic rats,[37] suggesting that increased lipogenesis in the nephrotic syndrome is not expressd constitutively, but is instead sensitive to dietary intake.

In addition to the lipid moieties, secretion of apolipoproteins is also increased.[28,35] A kinetic study has demonstrated a twofold increase in apo B synthesis in nephrotic rats.[38] The hepatic content of mRNAs encoding apolipoproteins is increased in the livers of nephrotic rats.[31,39]

Conflicting data on lipoprotein synthesis in patients with the nephrotic syndrome have been reported. Triglyceride synthesis was found to be increased in patients with the nephrotic syndrome.[40,41] However, Vega and Grundy[27] reported that the synthesis of VLDL was not increased in the four nephrotic patients that they studied. Warwick et al.[42] found that the synthesis of apo B was increased in nephrotic subjects and correlated with proteinuria in those patients in whom urinary protein loss exceeded 9 g/d. Apo B synthesis was not increased in patients with less proteinuria.

Mechanism of Increased Hepatic Synthesis of Lipoproteins in the Nephrotic Syndrome

The specific mechanism causing increased lipogenesis has not been elucidated; however, ample evidence suggests that reduced extracellular albumin concentration[43] and/or reduced extracellular π[44] in some way regulates apolipoprotein synthesis and lipogenesis by the hepatocyte. Increased plasma lipid levels in nephrotic patients, especially cholesterol levels, are inversely correlated with plasma protein or albumin concentration (Fig. 4-3). Many studies have shown a negative correlation between serum albumin and cholesterol concentrations in patients with the nephrotic

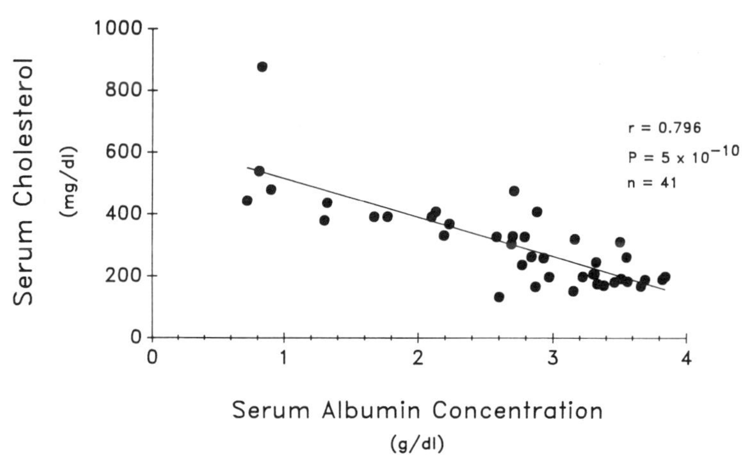

Fig. 4-3. Correlation between serum cholesterol concentration and serum albumin concentration in nephrotic patients. Each data point represents serum albumin concentration and is the mean of at least 10 measurements of each value on each patient obtained while patients were maintained on a constant diet.

syndrome[13,45–49] or a positive correlation between the reciprocal of serum albumin and cholesterol.[16] Similar relationships have been found between plasma albumin and plasma TGs and the phospholipids. Plasma cholesterol also correlates positively with urinary albumin[46] or protein excretion.[47,48] It may not be appropriate to infer causality from these relationships, however, since plasma albumin concentration as well as urinary albumin excretion are functions of both albumin synthesis[50] and altered glomerular permselectivity.[46] Like albumin, the plasma concentration of many other proteins that are synthesized in the liver is reduced in the nephrotic syndrome and the urinary excretion rate is increased.[2,51–59] A correlation between serum cholesterol and either the plasma concentration or urinary excretion rate could be as well constructed for any of these proteins as for albumin, even though their quantitative loss is much less.

Effect of π

The rate of albumin synthesis is increased[26,46,50,60] and plasma albumin decreased in the nephrotic syndrome. Marsh and Drabkin[4] proposed that hypoalbuminemia stimulated the hepatic synthesis of albumin and of other plasma proteins, including apolipoproteins. Many studies using cell culture, rats, or humans have been performed to test this hypothesis.

Extracellular albumin inhibited the secretion of apo B from cultured human hepatoblastoma-derived cells,[43] although there was no further increase in the effect of albumin above 10 g/L. Albumin concentrations of 30 g/L did not affect the level of apo B mRNA. This observation led the investigators to conclude that hepatic apo B synthesis was regulated under these circumstances at a step after the synthesis of mRNA, either by an increased rate of translation or by more efficient processing or secretion of the synthesized protein. In contrast, Yamauchi et al. reported that apo B mRNA levels could be modulated by increasing medium albumin concentration from 0 to 40 g/L in cultured hepatocytes.[44] It is possible that failure to detect an effect of π on hepatic apolipoprotein gene expression in some laboratories is a result of the experimental conditions used, and that extracellular π regulates apolipoprotein gene expression at several steps. In cultured rat hepatocytes, both bovine serum albumin (35 g/L) and dextran (81 kd, 35 g/L) decreased secretin of VLDL TGs into the medium.[61] This suggests that the inhibitory effect of albumin on lipogenesis may be a nonspecific effect of π rather than a specific effect of albumin.

Hyperlipidemia Is Not Linked to Increased Albumin Synthesis

It has been postulated that in the nephrotic syndrome, increased hepatic synthesis of lipids and lipoproteins occurs in parallel with, and as a result of, increased albumin synthesis as part of a generalized increase in the synthesis of all proteins secreted by the liver or as a result of liver hypertrophy.[4,62] Plasmapheresis, inducing hypoproteinemia in rats, also increases lipoprotein synthesis, both in vivo[33] and in perfused livers.[34] These observations are consistent with a hypothesis that increased synthesis of lipids and of hepatic proteins occurs in parallel. Not all apolipoproteins, however, exhibit an increased rate of synthesis in isolated perfused livers of nephrotic rats,[1] suggesting that their individual rates of synthesis are independently regulated. In addition, increased lipid levels may occur in nephrotic patients without an increased rate of albumin synthesis.[46] When proteinuria is reduced in nephrotic rats by treatment with enalapril, both plasma cholesterol and TG are also reduced, even though albumin synthesis rate,[63] and liver mass remain elevated.[64] Plasma lipid levels are also unrelated to albumin synthesis rate in normal rats.[26,63] Thus, no clear link can be established between hyperlipidemia and the rate of albumin synthesis, either in the rat or in humans. Increasing π in nephrotic rats by infusion of any of several macromolecules (albumin, polyvinylpyrrolidone, or dextran) causes a decrease in plasma lipids.[7] Albumin infusions also reduce cholesterol and phospholipids in patients with the nephrotic syndrome.[65] However, similar observations were made following dextran infusion, again suggesting that the effects of albumin may be in part nonspecific. Of the substances infused, only albumin increases LPL activity by binding fatty acids, indicating that the hypolipidemic effect is not due to a direct effect of albumin on lipolysis.[65]

Both plasma cholesterol and TG are increased in rats[66–68] and humans[69–72] with hereditary analbuminemia, a condition characterized by the virtual absence of albumin from the plasma and reduced π, but no urinary protein loss. Evidence supporting a link between decreased plasma π and hyperlipidemia in this disease is provided by the observations that plasma lipids are reduced in both analbuminemic rats[73] and humans,[69,74] as in animals[7] and patients[65] with the nephrotic syndrome, when oncotically active macromolecules (albumin or dextran) are infused intravenously. In conjunction with data showing that plasmapheresis causes rebound hyperlipidemia and increased lipid synthesis,[33] these observations support a hypothesis that reduced π contributes to hyperlipidemia in the nephrotic syndrome by stimulating hepatic lipogenesis. They do not, however, link increased hepatic lipogenesis with an increased rate of albumin synthesis, nor do they address alterations in lipoprotein catabolism that are also characteristic of the nephrotic syndrome.

It may be possible that synthesis of a group of plasma proteins, including apolipoproteins, is modulated either directly or indirectly by plasma π, although the specific mechanisms have yet to be elucidated. However the synthesis rate of these proteins is regulated quite independently of that of albumin.

LIPID CATABOLISM IN THE NEPHROTIC SYNDROME

The rate of catabolism of TG laden lipoproteins, CMs and VLDL, is greatly reduced in the nephrotic syndrome. The clearance of CM and VLDL is delayed in nephrotic rats,[5,26] the $t\frac{1}{2}$ being increased from 8 to 42 minutes

($P < .01$) for CMs (Fig. 4-4) and from 8 to 37 minutes ($P < 0.01$) for VLDL (Fig. 4-5). Garber et al.[3] found that LPL activity was reduced in nephrotic rats, and suggested that reduction in LPL activity was responsible for incomplete conversion of VLDL to IDL and LDL and reduced transfer of lipids to HDL. While it is possible that hypoalbuminemia could lead to reduced catabolism of circulating blood lipids by reducing either LPL or LCAT activity, as discussed at the beginning of this chapter, we find that rats with hereditary analbuminemia remove both CM and VLDL from the circulation normally[26] (Figs. 4-4 and 4-5), despite the absence of albumin from their plasma and despite marked reduction in post-heparin LPL activity. While albumin may play an important role both in facilitating the maturation of HDL and in optimizing LPL activity, clearly its presence in the plasma is not necessary for CMs or VLDL to be cleared from the blood at a normal rate.

When proteinuria is induced in analbuminemic animals, both serum cholesterol and TG concentrations increase significantly, and they develop a severe defect in the clearance of CM and VLDL that is no different from that which occurs in otherwise normal rats when the experimental nephrotic syndrome is induced. The defect in CM or VLDL clearance in the nephrotic

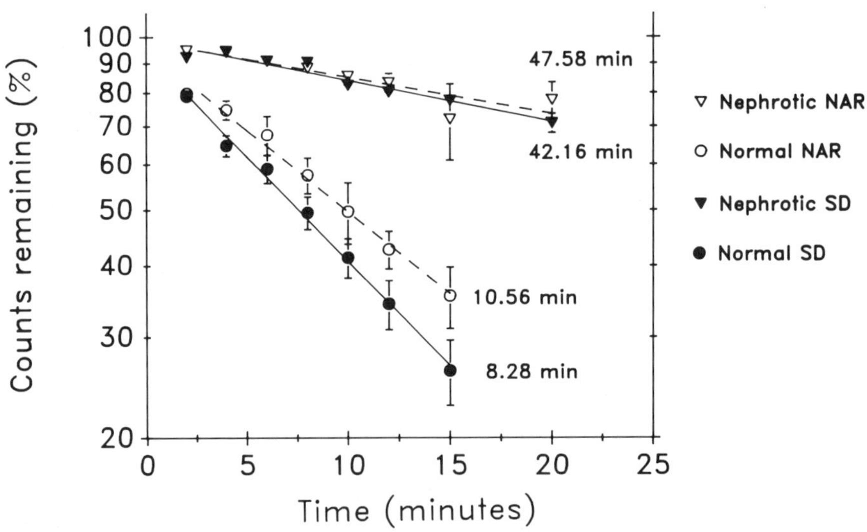

Fig. 4-4. Chylomicron clearance in normal and nephrotic Sprague-Dawley and Nagase analbuminemic rats. Chylomicron clearance was measured by the disappearance of ^{3}H-labeled CMs after intravenous injection. NAR, Nagase analbuminemic rat, represented by open symbols and broken lines. SD, Sprague-Dawley rat, represented by solid symbols and lines. Nephrotic rats are represented by inverted triangles and non-nephrotic rats by circles. The $t_{\frac{1}{2}}$ of each subgroup is indicated in the figure. Normal SD, N=6; normal NAR, N=6; nephrotic SD, N=8; nephrotic NAR, N=7. (From Davies et al.,[26] with permission.)

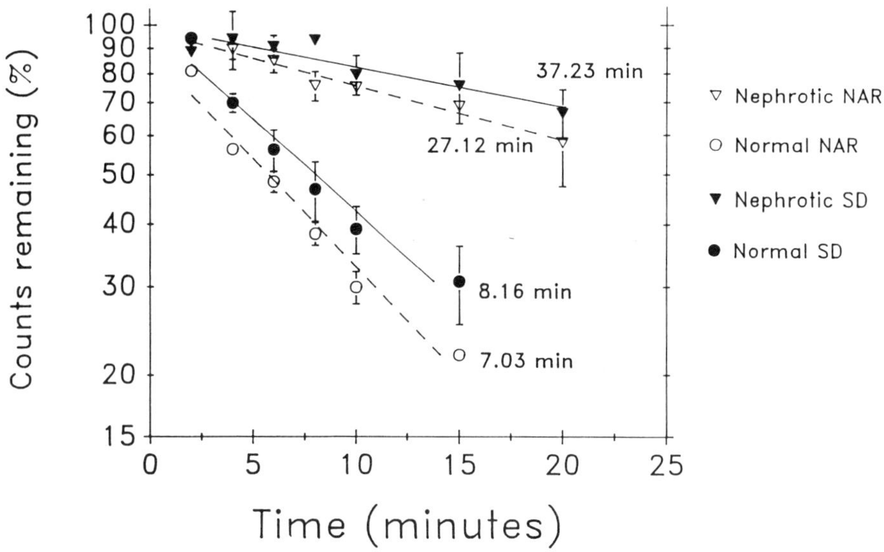

Fig. 4-5. VLDL clearance in normal and nephrotic Sprague-Dawley and Nagase analbuminemic rats. Abbreviations and symbols are as in Figure 4-4. (From Davies et al.,[26] with permission.)

syndrome is therefore not caused by alterations in albumin metabolism or plasma albumin concentration, but instead occurs following an alteration in glomerular permselectivity, possibly from the urinary loss of a substance(s) necessary for normal metabolism of VLDL and CMs.[26] Therefore, the nephrotic syndrome either alters lipoproteins themselves, so that their catabolic rate is reduced, or alters the pathways necessary for their metabolism (Figs. 4-1 and 4-2).

CM composition is altered in the nephrotic syndrome.[5] Remarkably, CMs obtained from nephrotic rats are cleared slightly faster than are normal CMs in control rats,[5] suggesting that delayed uptake of TG-rich particles occurs in the nephrotic environment. Thus, the alteration in CM or VLDL composition by itself is either insufficient to alter its catabolism or its structure is rapidly repaired in the normal environment. Furukawa et al. recently reported that HDL isolated from normal animals corrects defective catabolism of VLDL isolated from nephrotic rats by LPL in vitro[36] (Fig. 4-6). One would predict that preincubating normal VLDL with HDL obtained from nephrotic animals should show a reduced release of free fatty acids. This experiment, which has not been reported as yet, would identify a defect in HDL as a potential cause of reduced VLDL catabolism in the nephrotic syndrome.

HDL obtained from nephrotic animals is rich in apo A-I and deficient in apo E, apo A-IV,[3] and apo C-II.[32] These abnormalities could clearly contribute to defective clearance of VLDL or CMs, since apo C-II is necessary for full LPL activity and apo E serves as a target for a receptor protein in liver that

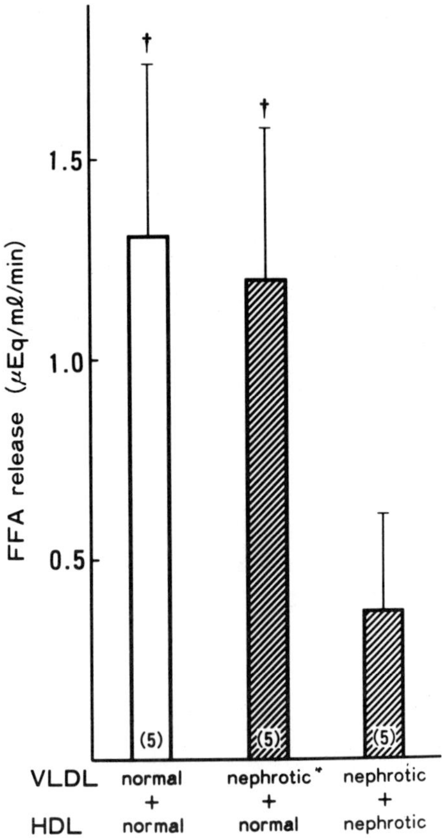

Fig. 4-6. Free fatty acid (FFA) release from VLDL triglycerides from either normal or nephrotic rats incubated with LPL. VLDL was preincubated with HDL from either normal or nephrotic rats. Numbers in parentheses indicate the number of donor rats that provided one VLDL sample. Results are expressed as mean ± SD. The dagger sign (†) represents significance against nephrotic VLDL + nephrotic HDL at $P < .01$. (From Furukawa et al.,[36] with permission.)

takes up CM and VLDL remnant particles.[3] Garber et al. postulated that this abnormality in HDL composition in the rat resulted from reduced transfer of these latter two apolipoproteins from VLDL to HDL.[3] The defect in VLDL metabolism was hypothesized to result from reduced LPL activity. Furukawa et al.,[36] however, found that defective VLDL catabolism was corrected in nephrotic rats by replacement with HDL from non-nephrotic rats, whereas HDL from nephrotic rats did not correct the abnormal VLDL clearance. These observations suggest that structural abnormalities in HDL may be a cause of, rather than a consequence of, abnormal VLDL catabolism.

LPL activity is reduced in both post-heparin plasma[3,26] and in adipose tissue in nephrotic rats.[5,75] In addition, the low levels of apo C-II in VLDL and CM in the nephrotic syndrome can contribute to a further reduction in effectiveness of LPL in vivo. However, LPL activity is extremely reduced both in post-heparin plasma[26] and in adipose tissue in the analbuminemic rat,[68] yet CM and VLDL clearances are normal in analbuminemic rats[26] (Figs. 4-4 and 4-5). These observations are especially striking considering that post-heparin LPL activity is nearly an order of magnitude greater in nephrotic Sprague-Dawley rats than in non-nephrotic analbuminemic rats.[26] Thus, it is unlikely that the defect in removal of CMs and VLDL from the blood results primarily from a decrease in LPL activity. Furthermore, Levy et al.[15] reported that uptake of TGs from either CMs or VLDL by tissues rich in LPL was reduced in nephrotic rats, suggesting that the reduced LPL activity was not the cause for decreased removal of these particles from the circulation in nephrosis. They suggested that defective removal was due to an abnormal composition of VLDL and CMs or that mechanisms of uptake were saturated due to an increase in the rate of synthesis of these particles. However, the rate of synthesis of these lipoproteins is not greatly increased in comparison to the increase in blood lipid levels.[27,36–38]

Studies in patients with the nephrotic syndrome have not been as detailed as in the rat; however, when comparable studies are evaluated, both species exhibit similar disturbances in lipid metabolism. The fractional turnover rate of TGs is reduced in nephrotic subjects compared with controls, and the $t_{\frac{1}{2}}$ of TG is prolonged from 4 to 11 hours in VLDL.[41] Not only is VLDL catabolism decreased, but the disappearance curve has an unusual shape, presumed to result from a delay in the conversion of VLDL to IDL.[27]

The delay in lipolysis in humans is proposed to be due to a decrease in LPL activity. Evidence supporting this hypothesis is that LPL activity is reduced in children with the nephrotic syndrome and increases after remission. Furthermore, there is a tight inverse correlation between LPL and the concentration of TGs in the VLDL fraction.[75] Although total plasma apo C-II is raised in nephrotic patients, the amount of this LPL cofactor per unit of VLDL is reduced by more than 50 percent since the total amount of VLDL is increased more than is total apo C-II.[30] These findings are similar to those described in rats.[28] Chylomicron metabolism has not been studied in human nephrosis. However, the fractional rate of disappearance of Intralipid (adminstered intravenously) is lower in nephrotic patients than in healthy controls.[76]

Although the delayed catabolism of TGs and the TG-rich lipoproteins has been very well established, much less is known about the catabolism of cholesterol and cholesterol-rich lipoproteins. LDL catabolism has been reported to be either normal[26] or reduced[42] in nephrotic rats.[38] However, LDL is not the principal cholesterol-bearing lipoprotein in rats; that function is carried out by HDL. The catabolism of the protein moieties, apo A-I and apo C, is reduced in HDL by 50 percent in nephrotic rats[32] irrespective of whether the HDL is derived from normal or nephrotic rats. Nothing is known, however, about the fate of cholesterol in HDL in the nephrotic syndrome.

Effect of Altered Glomerular Permselectivity on Lipid Metabolism

The measurement of the renal clearance of albumin is a crude measurement of glomerular permselectivity and directly reflects the underlying pathophysiologic condition causing proteinuria. Figures 4-3 and 4-7 illustrate the relationship between serum cholesterol and serum albumin and between serum cholesterol and the renal clearance of albumin, respectively. Although the correlation between serum cholesterol and both serum albumin (Fig. 4-3) and urinary albumin excretion (data not shown) are quite strong, the strongest correlation is between serum cholesterol concentration and the renal clearance of albumin (Fig. 4-7), a marker of altered glomerular permselectivity. Serum cholesterol was found to be independent of urinary albumin excretion or of the rate of albumin synthesis by multiple regression analysis.[46] Both serum cholesterol and TG concentration correlate instead with the renal clearance of albumin in patients with the nephrotic syndrome,[46] consistent with a hypothesis that altered glomerular permselectivity may affect lipid metabolism by the loss of any of a variety of substances that are retained by the normal glomerular basement membrane. This is further supported by the observation that reducing albuminuria by the use of angiotensin-converting enzyme inhibitors or by dietary protein restriction reduces blood lipid levels.[26,63,77]

It has been postulated that a factor needed to activate lipolysis is lost in the urine. Indeed, a glycosaminoglycan (GAG), identified as a low molecular weight heparan sulfate, has been isolated bound to alpha-1 acid glycoprotein

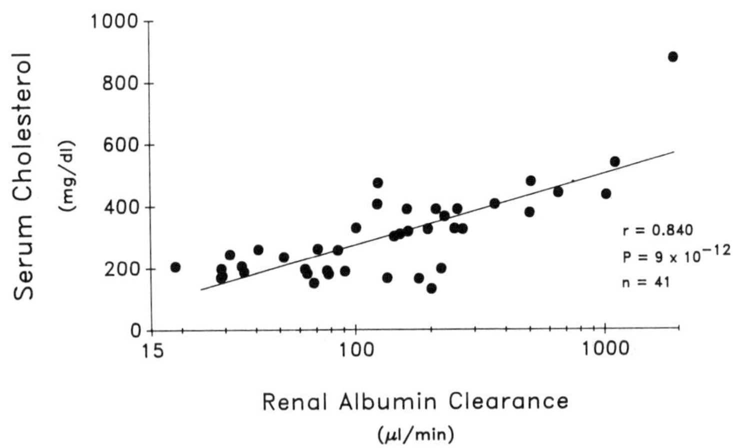

Fig. 4-7. Correlation between serum cholesterol concentration and the renal albumin clearance in nephrotic patients. Each data point represents a calculation of the renal clearance of albumin and is the mean of at least 10 measurements of both urinary albumin excretion and serum albumin concentration from each patient obtained while patients were maintained on a constant diet. Twenty-four-hour urine collections were obtained for 10 to 14 days consecutively.

from the urine of patients with the nephrotic syndrome.[78] This GAG increased the clearance of CMs when administered intravenously to nephrotic rats and increased lipolysis in vitro in the presence of apo C-II.[2,78,79] However, the urinary excretion of heparan sulphate is greater in normal subjects than in nephrotic humans and rats.[2,79] Thus, either a protein necessary for this GAG to affect lipolysis in vivo is lost in the urine, or the observation that heparan sulfate both stimulates lipolysis and corrects disordered CM clearance in nephrotic rats may be entirely coincidental and not germane to the diminished lipolysis in the nephrotic syndrome.

Let us digress for a moment about various substances lost in nephrotic urine that exert a putative affect on lipid metabolism.

Several macromolecules that affect lipid metabolism have been found in nephrotic urine in addition to heparan sulfate. One of these is HDL.[48] The term *lipoid nephrosis,* used to describe one entity that causes the nephrotic syndrome (specifically, minimal change nephrotic syndrome), is derived from lipid appearing in the urine and also visible in renal proximal tubular cells.[80] Lipiduria does occur in humans[81] and rats[82] with the nephrotic syndrome, but daily urinary HDL excretion is less than 1 percent of the total plasma HDL pool.[48] In addition, the HDL fraction that is reduced in the plasma of nephrotic patients is the 400 kd HDL_2 particle and not the 200 kd HDL_3 particle.[17] It is difficult to conceive of a mechanism invoking the urinary loss of HDL as a cause for its decreased plasma concentration that would reduce the levels of the larger but not the smaller of these lipoproteins. Furthermore, HDL levels are increased in rats with the nephrotic syndrome.

HDL maturation is dependent in part on the activity of LCAT.[10,12] This enzyme has also been found in the urine of nephrotic rats.[83] While HDL maturation may be reduced because of a reduced rate of cholesterol esterification by LCAT,[11,15] it is unlikely that the urinary loss of LCAT contributes to the reduced plasma levels of this enzyme.[84] Apo C-II is lost in the urine of nephrotic patients, but there is no correlation between urinary apo C-II excretion and plasma apo C-II levels.[30] Furthermore, the absolute levels of apo C-II are not reduced in the nephrotic syndrome.[30]

Thus, although it is likely that deranged lipid catabolism results in part from the urinary loss of a substance or of a group of substances,[26,46] hyperlipidemia is not likely to result from the urinary loss of any of the putative liporegulatory substances isolated from the urine of nephrotic patients or animals to date.

TREATMENT OF HYPERLIPIDEMIA IN THE NEPHROTIC SYNDROME

Hyperlipidemia of the nephrotic syndrome may be severe and is characterized by a high LDL to HDL cholesterol ratio. This pattern is associated with accelerated atherosclerosis in other clinical situations, and there is no reason to believe that hyperlipidemia of the nephrotic syndrome would not have similar consequences should this metabolic disorder persist. Furthermore, LDL cholesterol is taken up by the mesangium of the glomerulus, causing progression of renal vascular resistance by enhancing thromboxane generation[87] in rats. Reducing blood cholesterol levels may therefore have a beneficial effect on the course of renal disease, although this has not been demonstrated in humans.

If the primary mechanism causing the nephrotic syndrome can be directly treated, with the use of glucocorticoids for the treatment of minimal change nephrotic syndrome for example, then treatment of hyperlipidemia is of limited clinical utility. If the primary disease process causing the nephrotic syndrome is untreatable, and proteinuria is anticipated to be prolonged, specific pharmacologic therapy should be directed at reducing markedly elevated blood lipid levels. Pharmacologic agents such as probucol,[88,89] gemfibrozil,[90] or HMG-CoA reductase inhibitors[26,91–93] may ultimately prove useful in reducing blood lipid levels in these patients, but use of these agents may engender some morbidity, such as rhabdomyolysis with lovastatin,[94] especially in combination with gemfibrozil.[95] It may be useful to obtain a baseline measurement of creatinine kinase prior to therapy with these agents. Plasma lipid levels should be monitored to document efficacy of treatment, and patients should be followed for evidence of rhabdomyolysis. However, therapy of hyperlipidemia of the nephrotic syndrome should first be directed at reducing the increased renal clearance of macromolecules by the use of dietary protein restriction[96,97] and/or administration of angiotensin-converting enzyme inhibitors.[98]

EFFECT OF TREATMENT OF HYPERLIPIDEMIA OF THE NEPHROTIC SYNDROME ON PROTEINURIA

Cholestyramine ameliorates hypercholesterolemia and hypertriglyceridemia in rats rendered nephrotic with PAN. Proteinuria is reduced in the treated group at an early stage of disease, and treated animals do not develop progressive loss of renal function characteristic of this disease model.[99] Similarly, probucol also ameliorates hypercholesterolemia, hypertriglyceridemia, and proteinuria at an early stage in this disease model.[100] Pravastatin, an HMG-CoA reductase inhibitor, ameliorates hypertriglyceridemia but not hypercholesterolemia or proteinuria in this disease model.[101] Thus, the protective effect of lipid-lowering agents is not confined to one class of drugs, but follows instead from reduction in blood cholesterol levels. These observations have not been confirmed in humans. However, in a study without contemporaneous time controls, Rabelink et al. reported that albuminuria decreased and serum albumin and total protein increased in a group of nephrotic patients treated with an HMG-CoA reductase inhibitor for 48 weeks.[102]

ACKNOWLEDGMENT

This work was supported in part by the research service of the United States Department of Veterans Affairs; in part by a grant from the National Institutes of Health (1-RO1 DK 42297-01 PTHA), Institute of Diabetes and Digestive and Kidney Diseases; and in part by a grant from the Netherlands Heart Foundation (no. 88.223).

REFERENCES

1. Marsh JB: Lipoprotein metabolism in experimental nephrosis. J Lipid Res 25:1619, 1984
2. Staprans I, Felts JM, Couser WG: Glycosaminoglycans and chylomicron metabolism in control and nephrotic rats. Metabolism 36:496, 1987
3. Garber DW, Gottlieb BA, Marsh JB, Sparks CE: Catabolism of very low density lipo-proteins in experimental nephrosis. J Clin Invest 74:1375, 1984
4. Marsh JB, Drabkin, DL: Experimental reconstruction of metabolic pattern of lipid nephrosis: key role of hepatic protein synthesis in hyperlipemia. Metabolism 9:946, 1960
5. Levy E, Ziv E, Bar-On H, Shafrir E: Experimental nephrotic syndrome: removal and tissue distribution of chylomicrons and very-low-density lipoproteins of normal and nephrotic origin. Biochim Biophys Acta 1043:259, 1990
6. Eckel RH: Lipoprotein lipase: a multifunctional enzyme relevant to common metabolic diseases. N Engl J Med 320:1044, 1989
7. Allen JC, Baxter JH, Goodman HC: Effects of dextran, polyvinylpyrrolidone and gamma globulin on the hyperlipidemia of experimental nephrosis. J Clin Invest 40:499, 1961
8. Schaefer EJ, Levy RI: Pathogenesis and management of lipoprotein disorders. N Engl J Med 312:1300, 1985
9. Havel RJ: Lipid transport function of lipoproteins in blood plasma. Am J Physiol 253: E1, 1987
10. Eisenberg S: High density lipoprotein metabolism. J Lipid Res 25:1017, 1984
11. Cohen SL, Cramp DG, Lewis AD, Tickner TR: The mechanism of hyperlipidemia in nephrotic syndrome: role of low albumin and the LCAT reaction. Clin Chim Acta 104:393, 1980
12. Tall AR: Plasma high density lipoproteins: metabolism and relationship to atherogenesis. J Clin Invest 86:379, 1990
13. Conwill DE, Granger DN, Cook BH et al: The effect of serum oncotic pressure on serum cholesterol levels: a study in "normal" and nephrotic subjects. South Med J 70:456, 1977
14. Baxter, JH, Goodman HC, Havel RJ: Serum lipid and lipoprotein alterations in nephrosis. J Clin Invest 39:455, 1960
15. Nayak SS, Bhaskaranand N, Kamath KS et al: Serum apolipoproteins A and B, lecithin: cholesterol acyl transferase activities and urinary cholesterol levels in nephrotic syndrome patients before and during steroid treatment. Nephron 54:234, 1990

16. Gherardi E, Rota E, Calandra S et al: Relationship among the concentrations of serum lipoproteins and changes in their chemical composition in patients with untreated nephrotic syndrome. Eur J Clin Invest 7:563, 1977
17. Muls E, Rosseneu M, Daneels R et al: Lipoprotein distribution and composition in the human nephrotic syndrome. Atherosclerosis 54:225, 1985
18. Solkolovskaya IV, Nikiforova NV: High-density lipoprotein cholesterol in patients with untreated and treated nephrotic syndrome. Nephron 37:49, 1984
19. Karádi I, Romics L, Pálos G et al: Lp(a) lipoprotein concentration in serum of patients with heavy proteinuria of different origin. Clin Chem 35:2121, 1989
20. Scanu AM, Fless GM: Lipoprotein (a): heterogeneity and biological relevance. J Clin Invest 85:1709, 1990
21. Kostner GM, Avogaro P, Cazzolato G et al: Lipoprotein Lp(a) and the risk for myocardial infarction. Atherosclerosis 38:51, 1981
22. Utermann G: The mysteries of lipoprotein(a). Science 246:904, 1989
23. Boerwinkle E, Menzel HJ, Kraft HG, Utermann G: Genetics of the quantitative Lp(a) lipoprotein trait. III. Contribution of Lp(a) glycoprotein phenotypes to normal lipid variation. Hum Genet 82:73, 1989
24. Maeda S, Abe A, Seishima M et al: Transient changes of serum lipoprotein(a) as an acute phase protein. Atherosclerosis 78:145, 1989
25. Oschry Y, Eisenberg S: Rat plasma lipoproteins: re-evaluation of a lipoprotein system in an animal devoid of cholesterol ester transfer activity. J Lipid Res 23:1099, 1982
26. Davies RW, Staprans I, Hutchison FN, Kaysen GA: Proteinuria, not altered albumin metabolism, affects hyperlipidemia in the nephrotic rat. J Clin Invest 86:600, 1990
27. Vega GL, Grundy SM: Lovastatin therapy in nephrotic hyperlipidemia: effects on lipoprotein metabolism. Kidney Int 33:1160, 1988
28. Marsh JB, Sparks CE: Hepatic secretion of lipoproteins in the rat and the effect of experimental nephrosis. J Clin Invest 64:1229, 1979
29. Levy E, Ziv E, Bar-On H, Shafrir E: Chylomicron synthesis in experimental nephrotic syndrome. Biochim Biophys Acta 1005:20, 1989
30. Kashyap ML, Srivastava LS, Hynd BA et al: Apolipoprotein C II and lipoprotein lipase in human nephrotic syndrome. Atherosclerosis 35:29, 1980
31. Marshall JF, Apostolopoulos JJ, Brack CM, Howlett GJ: Regulation of apolipoprotein gene expression and plasma high-density lipoprotein composition in experimental nephrosis. Biochim Biophys Acta 1042:271, 1990
32. Sparks CE, Tennenberg SD, Marsh JB: Catabolism of the apolipoproteins of HDL in control and nephrotic rats. Biochim Biophys Acta 665:8, 1981
33. Shafrir E, Brenner T: Lipoprotein lipid and protein synthesis in experimental nephrosis and plasmapheresis. I: Studies in rat in vivo. Lipids 14:695, 1979
34. Brenner T, Shafrir E: Lipoprotein lipid and protein synthesis in experimental nephrosis and plasmapheresis. II. Perfused rat liver. Lipids 15:637, 1980
35. Calandra S, Gherardi F, Fainaru M et al: Secretion of lipoproteins, apolipoprotein A-I and apolipoprotein E by isolated and perfused liver of rat with experimental nephrotic syndrome. Biochim Biophys Acta 665:331, 1981
36. Furukawa S, Hirano T, Mamo JCL et al: Catabolic defect of triglyceride is associated with abnormal very-low-density lipoprotein in experimental nephrosis. Metabolism 39:101, 1990
37. Bar-Tana J, Rose-Kayn G, Frenkel B: Hypolipidemic effect of β,β'-methyl-substituted hexadecanedioic acid (MEDICA 16) in normal and nephrotic rats. J Lipid Res 29:431, 1988

38. Joven J, Masana L, Villabona C et al: Low density lipoprotein metabolism in rats with puromycin aminonucleoside-induced nephrotic syndrome. Metabolism 38:491, 1989
39. Tarugi P, Calandra S, Chan L: Changes in apolipoprotein A-I mRNA level in the liver of rats with experimental nephrotic syndrome. Biochim Biophys Acta 868:51, 1986
40. Kekki M, Nikkilä EA: Plasma triglyceride metabolism in the adult nephrotic syndrome. Eur J Clin Invest 1:345, 1971
41. McKenzie IFC, Nestel PJ: Studies on the turnover of triglyceride and esterified cholesterol in subjects with the nephrotic syndrome. J Clin Invest 47:1685, 1968
42. Warwick GL, Caslake MJ, Boulton-Jones JM et al: Low-density lipoprotein metabolism in the nephrotic syndrome. Metabolism 39:187, 1990
43. Pullinger CR, North JD, Teng BB et al: The apolipoprotein B gene is constitutively expressed in HepG2 cells: regulation of secretion by oleic acid, albumin, and insulin, and measurement of the mRNA half-life. J Lipid Res 30:1065, 1989
44. Yamauchi A, Yamamoto S, Fukuhara Y et al: Oncotic pressure regulates the levels of albumin (Alb) mRNA and apolipoprotein B (ApoB) mRNA in cultured rat hepatoma cells (H4IIE), abstracted. Kidney Int 35:441, 1989
45. Thomas EM, Rosenblum AH, Lander HB, Fisher R: Relationship between blood lipid and blood protein levels in the nephrotic syndrome. Am J Dis Child 81:207, 1951
46. Kaysen GA, Gambertoglio J, Felts J, Hutchison FN: Albumin synthesis, albuminuria and hyperlipidemia in nephrotic patients. Kidney Int 31:1368, 1987
47. Appel GB, Blum CB, Chien S et al: The hyperlipidemia of the nephrotic syndrome. N Engl J Med 312:1544, 1985
48. Jüngst D, Caselmann WH, Kutschera P, Weisweiler P: Relation of hyperlipidemia in serum and loss of high density lipoproteins in urine in the nephrotic syndrome. Clin Chim Acta 168:159, 1987
49. Michaeli J, Bar-On H, Shafrir E: Lipoprotein profiles in a heterogeneous group of patients with nephrotic syndrome. Israel J Med Sci 17:1001, 1981
50. Kaysen GA, Kirkpatrick WG, Couser WG: Albumin homeostasis in the nephrotic rat: nutritional considerations. Am J Physiol 247:F192, 1984
51. Jensen H, Bro-Jorgensen K, Jarnum SM: Transferrin metabolism in the nephrotic syndrome and in protein-losing gastroenteropathy. Scand J Clin Lab Invest 21:293, 1968
52. Ellis D: Anemia in the course of the nephrotic syndrome secondary to transferrin depletion. J Pediatr 90:953, 1977
53. Hancock DE, Onstad JW, Wolf PL: Transferrin loss into the urine with hypochromic, microcytic anemia. Am J Clin Pathol 65:72, 1976
54. Cartwright GE, Gubler CJ, Wintrobe MM: Studies on copper metabolism: copper and iron metabolism in the nephrotic syndrome. J Clin Invest 33:685, 1954
55. Goldstein DA, Haldimann B, Sherman D et al: Vitamin D metabolites and calcium metabolism in patients with nephrotic syndrome and normal renal function. J Clin Endocrinol Metab 53:116, 1981
56. Emerson K, Jr, Beckman WW: Calcium metabolism in nephrosis. I. A description of an abnormality in calcium metabolism in children with nephrosis. J Clin Invest 24:564, 1945

57. Haddad JG, Jr, Walgate J: Radioimmunoassay of the binding protein for vitamin D and its metabolites in human serum: concentrations in normal subjects and patients with disorders of mineral homeostasis. J Clin Invest 58:1217, 1976
58. Barragry JM, France MW, Carter ND: Vitamin D metabolism in nephrotic syndrome. Lancet 2:629, 1977
59. Staprans I, Anderson CD, Lurz FW, Felts JM: Separation of a lipoprotein lipase cofactor from the α_1 acid glycoprotein fraction from urine of nephrotic patients. Biochim Biophys Acta 617:514, 1980
60. Rothschild MA, Oratz M, Schreiber SS: Albumin synthesis. p. 249. In Javitt NB (ed): Liver and Biliary Tract Physiology. I. International Review of Physiology. Vol. 21. University Park Press, Baltimore, 1980
61. Davis RA, Engelhorn SC, Weinstein DB, Steinber D: Very low density lipoprotein secretion by cultured rat hepatocytes. J Biol Chem 255:2039, 1980
62. Goldberg ACRK, Oliveira HCF, Quintao ECR, McNamara DJ: Increased hepatic cholesterol production due to liver hypertrophy in rat experimental nephrosis. Biochim Biophys Acta 710:71, 1982
63. Kaysen GA, Davies RW: Reduction in proteinuria attenuates hyperlipidemia in the nephrotic syndrome. J Am Soc Nephrol 1:S75, 1991
64. Kaysen GA, Davies RW, Hutchison FN: Effect of dietary protein intake and ACE inhibition in Heymann nephritis. Kidney Int 36:S154, 1989
65. Baxter JH, Goodman HC, Allen JC: Effects of infusions of serum albumin on serum lipids and lipoproteins in nephrosis. J Clin Invest 40:490, 1961
66. Takahashi M, Kusumi K, Shumiya S, Nagase S: Plasma lipid concentrations and enzyme activities in Nagase analbuminemia rats (NAR). Exp Animal 32:39, 1983
67. Ando S, Kon K, Tanaka Y et al: Characterization of hyperlipidemia in Nagase analbuminemic rat (NAR). Biochemistry 87:1859, 1980
68. Kikuchi H, Tamura S, Nagase S, Tsuiki S: Hypertriacylglycerolemia and adipose tissue lipoprotein lipase activity in the Nagase analbuminemic rat. Biochim Biophys Acta 744:165, 1983
69. Waldmann TA, Gordon RS, Jr, Rosse W: Studies on the metabolism of the serum proteins and lipids in a patient with analbuminemia. Am J Med 37:960, 1964
70. Baldo-Enzi G, Baiocchi MR, Vigna G et al: Analbuminemia: a natural model of metabolic compensatory systems. J Inherited Metab Dis 10:317, 1987
71. Dammacco F, Miglietta A, D'Addabbo A et al: Analbuminemia: a report of a case and review of the literature. Vox Sang 39:153, 1980
72. Russi E, Weigand K: Analbuminemia. Klin Wochenschr 61:541, 1983
73. Edelstein D, Arbeeny C, Chowdhury JR: Effect of albumin and dextran on the hyperlipidemia in analbuminemic rats, abstracted. Arteriosclerosis 7:535, 1987
74. Cormode EJ, Lyster MD, Israels S: Analbuminemia in a neonate. J Pediatr 86:862, 1975
75. Yamada M, Matsuda I: Lipoprotein lipase in clinical and experimental nephrosis. Clin Chim Acta 30:787, 1970
76. Chan MK, Persaud JW, Ramdial L et al: Hyperlipidaemia in untreated nephrotic syndrome, increased production or decreased removal? Clin Chim Acta 117:317, 1981
77. Kaysen GA, Don B, Schambelan M: Proteinuria, albumin synthesis and hyperlipidemia in the nephrotic syndrome. Nephrol Dial Transplant 6:141, 1991

78. Staprans I, Felts JM: The effect of α1-acid glycoprotein (orosomucoid) on triglyceride metabolism in the nephrotic syndrome. Biochem Biophys Res Commun 79:1272, 1977
79. Staprans I, Garon SJ, Hopper J, Felts JM: Characterization of glycosaminoglycans in urine from patients with nephrotic syndrome and control subjects and their effects on lipoprotein lipase. Biochim Biophys Acta 678:414, 1981
80. McCoy RC, Tischer CC: The pathology of primary glomerular diseases. p. 585. In Early LE, Gottschalk CW (eds): Strauss and Welt's Diseases of the Kidney. 3rd Ed. Little, Brown, Boston, 1979
81. Shore VG, Forte T, Licht H, Lewis SB: Serum and urinary lipoproteins in the human nephrotic syndrome: evidence for renal catabolism of lipoproteins. Metabolism 31:258, 1982
82. Gherardi E, Calandra S: Plasma and urinary lipids and lipoproteins during the development of nephrotic syndrome induced in the rat by puromycin aminonucleoside. Biochim Biophys Acta 710:188, 1981
83. Gherardi E, Vecchia L, Calandra S: Experimental nephrotic syndrome in the rat induced by puromycin aminonucleoside. Plasma and urinary lipoproteins. Exp Mol Pathol 32:128, 1980
84. Sestak TL, Alavi N, Subbaiah PV: Plasma lipids and acyltransferase activities in experimental nephrotic syndrome. Kidney Int 36:240, 1989
85. Moorhead JF, Wheeler DC, Varghese Z: Glomerular structures and lipids in progressive renal disease. Am J Med 87:12, 1989
86. Keane WF, Kasiske BL, O'Donnell MP: Lipids and progressive glomerulosclerosis: a model analogous to atherosclerosis. Am J Nephrol 8:261, 1988
87. Kaplan R, Aynedjian HS, Bank N, Schlondorff D: Cholesterol feeding causes renal vasoconstriction via oxidized lipoprotein activation of thromboxane, abstracted. Kidney Int 37:371A, 1990
88. Iida H, Izumino K, Asaka M: Effect of probucol on hyperlipidemia in patients with nephrotic syndrome. Nephron 47:280, 1987
89. Valeri A, Gelfand J, Blum C, Appel GB: Treatment of the hyperlipidemia of the nephrotic syndrome: a controlled trial. Am J Kidney Dis 8:388, 1986
90. Groggel GC, Cheung AK, Ellis-Benigni K, Wilson DE: Treatment of nephrotic hyperlipoproteinemia with gemfibrozil. Kidney Int 36:266, 1989
91. Golper TA, Illingworth DR, Morris CD, Bennett WM: Lovastatin in the treatment of multifactorial hyperlipidemia associated with proteinuria. Am J Kidney Dis 13:312, 1989
92. Kasiske BL, Velosa JA, Halstenson CE et al: The effects of lovastatin in hyperlipidemic patients with the nephrotic syndrome. Am J Kidney Dis 15:8, 1990
93. Rabelink AJ, Hené RJ, Erkelens DW et al: Effects of simvastatin and cholestyramine on lipoprotein profile in hyperlipidaemia of nephrotic syndrome. Lancet 2:1335, 1988
94. Mccorpier CL, Jones PH, Suki WN et al: Rhabdomyolysis and renal injury with lovastatin use. Report of two cases in cardiac transplant recipients. JAMA 260:239, 1988
95. Marais GE, Larson KK: Rhabdomyolysis and acute renal failure induced by combination lovastatin and gemfibrozil therapy. Ann Intern Med 112:228, 1990
96. Kaysen GA, Gambertoglio J, Jiminez I et al: Effect of dietary protein intake on albumin homeostasis in nephrotic patients. Kidney Int 29:572, 1986
97. Rosenberg ME, Swanson JE, Thomas BL, Hostetter TH: Glomerular and hor-

monal responses to dietary protein intake in human renal disease. Am J Physiol 253:F1083, 1987
98. Don BR, Kaysen GA, Hutchison FN, Schambelan M: The effect of angiotensin converting enzyme inhibition and dietary protein restriction in the treatment of proteinuria. Am J Kidney Dis 17:10, 1991
99. Diamond JR, Hanchak NA, McCarter MD, Karnovsky MJ: Cholestyramine resin ameliorates chronic aminonucleoside nephrosis. Am J Clin Nutr 51:606, 1990
100. Hirano T, Mamo JCL, Ohnami K, Sugisaki T: Treatment of hyperlipidemia with probucol reduces proteinuria in puromycin aminonucleoside-induced nephrotic rats. Nephron 54:358, 1990
101. Hirano T, Komuro F, Furukawa S et al: Effect of pravastatin sodium, a new inhibitor of 3-hydroxy-3-methylglutaryl coenzyme A reductase, on very-low-density lipoprotein composition and kinetics in hyperlipidemia associated with experimental nephrosis. Metabolism 39:605, 1990
102. Rabelink AJ, Hene RJ, Erkelens DW et al: Partial remission of nephrotic syndrome in patients on long-term simvastatin. Lancet 335:1045, 1990

5

Non-Cyclooxygenase Lipid Mediators and Regulation of Renal Function

Kamal F. Badr

INTRODUCTION

The metabolism of arachidonic acid (AA) by the cyclooxygenase enzyme results in the formation of prostaglandins and thromboxanes, which have long been known to play a central role in the control of organ function and homeostasis. In the kidney, these eicosanoids are involved in the regulation of renal vascular tone and tubule function under normal conditions, and in mediating alterations in glomerular function during injury. Over the past decade, increasing attention has been directed at alternative pathways of arachidonate oxygenation, namely, those catalyzed by the lipoxygenases (LOs) and, more recently, by the cytochrome P-450 enzyme system. While LO products are generated primarily from inflammatory cells, and thus participate in the mediation of renal inflammatory injury, cytochrome P-450–AA derivatives are synthesized in the normal kidney and appear to regulate renal vascular tone and salt and water transport. We review recent advances in our understanding of the roles of non-cyclooxygenase–derived lipid mediators in the regulation of renal function under normal and pathologic conditions. Of necessity, such a review also includes the prominent renal actions of platelet-activating factor, a glycerophosphocholine.

THE LIPOXYGENASE PATHWAY

Chemistry

In activated leukocytes and platelets, AA is oxygenated via 5-, 12-, and 15-LO enzymes into biologically active metabolites[1] (Figs. 5-1 and 5-2). In these reactions, hydroperoxy groups are introduced onto the fatty acid backbone at C5, C12, and C15, respectively. Following the formation of 5-hydroperoxyeicosatetraenoic acid (5-HPETE), 5-LO also catalyzes the formation, from 5-HPETE, of a 5,6-epoxy intermediate, leukotriene A_4 (LTA_4). The metabolic fate of LTA_4 is depicted in Fig. 2-1, and the resulting LTs (B_4, C_4, D_4, and E_4) exert important biologic effects in the kidney.[2–6] Note that the formation of LTB_4 requires the enzyme LTA_4 epoxide hydrolase, and that of the peptido-LTs (C_4, D_4, and E_4) requires a glutathione transferase (LTC_4 synthase). 5-Lipoxygenase and LTA_4 hydrolase have been recently cloned and sequenced from human and rat leukocytes.[7,8] While these enzymes are all present in leukocytes, it is now well established that the biosynthesis of LTs can take place in erythrocytes,[9] endothelial cells,[10] and platelets[11,12] by transcellular metabolism of leukocyte-generated LTA_4.[9–12] This pathway for LT generation may be particularly important in the glomerulus (see below).

Sequential actions by 5- and 15-LO enzymes result in the formation of epoxytetraene intermediates, leading to the generation of trihydroxy AA derivatives, the lipoxins (LXs)[1] (Fig. 2-2). These include LXA_4, LXB_4, and the recently identified LXA_4 isomer, 7-*cis,* 11-*trans*-LXA_4.[1,13] 15-Lipoxygenase may also act directly on non-oxygenated AA, yielding 15-

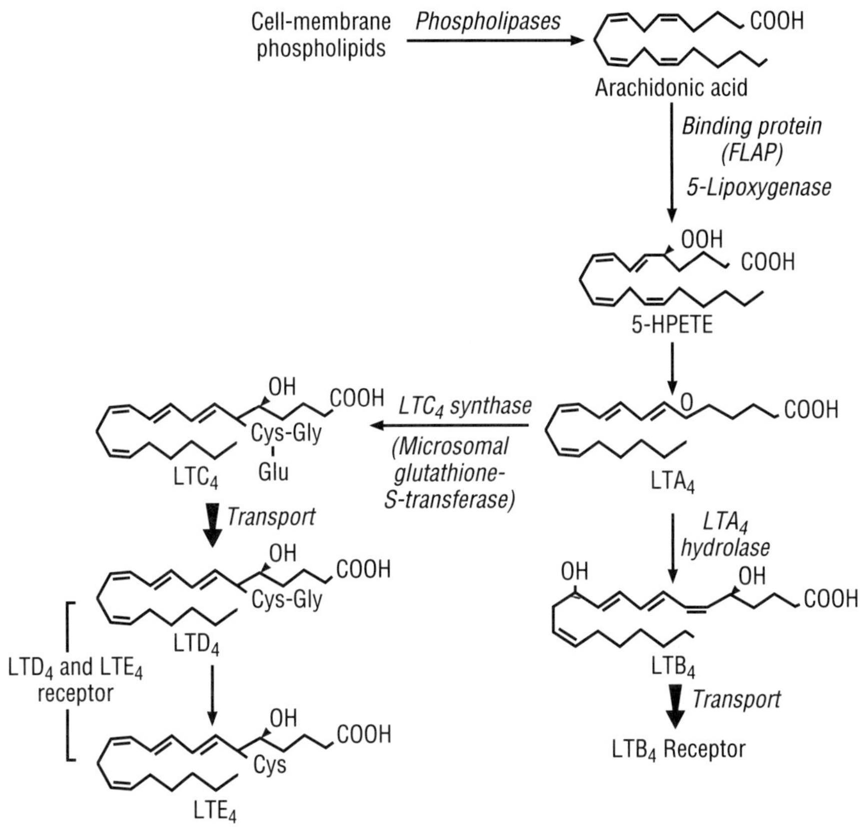

Fig. 5-1. 5-Lipoxygenase pathway reactions leading to the formation of leukotrienes. FLAP, 5-lipoxygenase activating protein; transport, specific transport processes for extracellular export.

HPETE and 15-hydroxyeicosatetaraenoic acid (15-HETE). The 12-LO reaction is the prominent oxygenation reaction of AA in platelets, leading to the formation of 12-HETE. Recently, Serhan and Sheppard demonstrated that 12-LO is also capable of generating LXs through transcellular metabolism of leukocyte-derived LTA_4.[14]

Renal Sources and Nephronal Distribution of Lipoxygenase Products

The major sources of LO products derived in the kidney are infiltrating leukocytes or resident cells of macrophage/monocyte origin. Increasing evidence, however, suggests that intrinsic renal cells can generate LTs and LXs, likely through the transcellular metabolism of intermediates, particularly

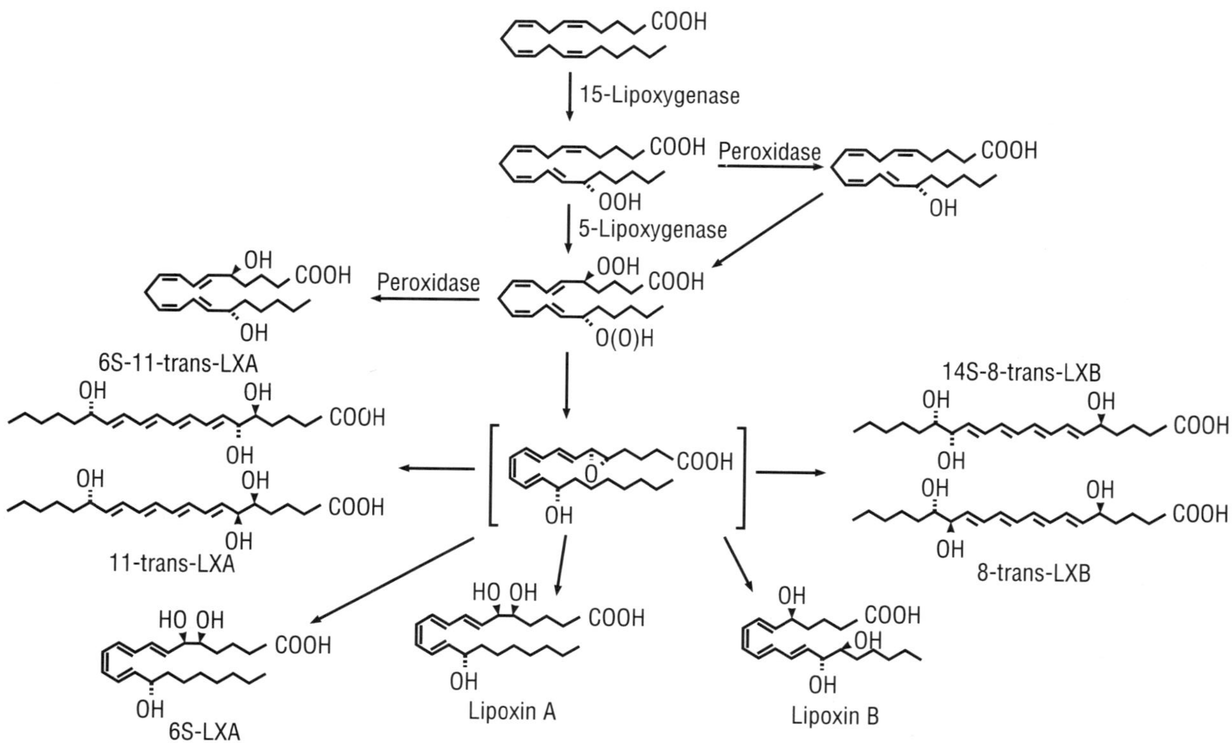

Fig. 5-2. Enzymatic pathways of LX formation in polymorphonuclear neutrophil leukocytes.

LTA_4.[15–23] Human and rat glomeruli can generate 12- and 15-HETE,[18,19] although the cells of origin are unclear and small amounts of 5-HETE were detected from rat glomerular preparations.[19] In studies by Cattell et al.[16] and Lefkowith et al.,[24] LTB_4 was detected in supernatants of normal rat glomeruli and its synthesis could be markedly diminished by maneuvers that depleted glomeruli of resident macrophages, such as irradiation or fatty acid deficiency.[25] In addition, 5-, 12-, and 15-HETE were detected from pig glomeruli and their structural identity was confirmed by mass spectrometry.[15] Glomeruli subjected to immune injury release LTB_4,[26–29] and LTB_4 generation is again suppressed by resident macrophage depletion.[29] Synthesis of peptido-LTs by inflamed glomeruli has also been demonstrated.[30] The same applies to the immune-injured isolated perfused kidney, which has been reported to generate LTC_4 and LTD_4.[21] Cultured glomerular epithelial[19] and mesangial[20] cells also synthesize 12-HETE. The glomerular synthesis of LXs is uncharacterized, but rat mesangial cells have been recently reported to generate LXA_4 when provided with LTA_4 as substrate, thereby providing a potential intraglomerular source of LXs during inflammatory reactions.[22] In nonglomerular tissue, 12-HETE production has been reported from rat cortical tubules and epithelial cells[19] and 12- and 15-HETE production from rabbit medulla.[31]

In summary, glomeruli are capable of generating LTs, but the exact cellular origin of 5-LO products remains uncertain. The glomerular biosynthesis of all LO products is distinctly stimulated following in vitro or in vivo cellular activation, particularly by immune injury.

Renal Metabolism of Leukotrienes

LTC_4 is converted rapidly to LTD_4 and LTE_4 in rat glomeruli and papilla by sequential peptide cleavage.[32] The reaction is catalyzed by dehydropeptidase-I, which has been solubilized from rat kidney microsomes.[33] *N*-Acytelation of LTE_4 leads to loss of biologic activity. The renal metabolism of LTB_4 proceeds via conversion to the 6-*trans*-isomer (biologically inactive), as detected in rat kidney homogenates,[34] or by primary reduction of double bonds leading to the formation of 5,12-diHETE[35] in mesangial cells, the biologic activity of which is undefined. The renal metabolism of LXs is unexplored.

Renal Receptors and Intracellular Signals for Lipoxygenase Products

Binding studies of radiolabeled LTC_4 to whole glomeruli revealed the presence of specific low-affinity binding sites.[36] LTC_4 binding was also described in human glomerular epithelial cells, with similarly low affinity (kd = 220 nmol/L.[37] Subsequent studies,[38] however, have confirmed that LTC_4 binding sites in general are likely not representative of surface receptors, but involve a cytosolic site linked to glutathione-S-transferase activity. The principal LT for which specific binding and true membrane-bound receptors have been identified in extrarenal tissue is LTD_4.[39–42] Recently, specific high-affinity LTD_4 receptors have been identified on rat and human glomerular mesangial cells.[43,44] The dissociation constants in both species are remarkably close (10 and 12 nmol/L, respectively), as is the rank order of potency for binding competition by other LTs ($LTD_4 > LTE_4 = LTC_4 >> LTB_4$). Of interest, a crucial stereochemical requirement for LTD_4 binding appears to be the S,R orientation of the polar substituents at C5 and C6, such that 5R,6S-LTD_4 competes for binding with very low potency (= LTB_4).[43] LXA_4, which possesses similar stereochemistry at these carbons, competes well for ^{3}H-LTD_4 binding to rat mesangial cells and elicits weak agonist activity (Figs. 5-3 and 5-4).[45]

Intracellular signaling for LTD_4 in human and rat mesangial cells involves receptor activated phosphoinositolbisphosphate (PIP_2) hydrolysis, release of inositol phosphates, and increased intracellular calcium concentrations[43–45] (Fig. 5-5). Other responses include amiloride-sensitive intracellular alkalinization and mitogenesis.[43,44] LTC_4 is also mitogenic to rat glomerular epithelial cells and appears to also activate protein kinase C and intracellular alkalinization in those cells.[37,46]

Stimuli for Lipoxygenase Product Release in the Kidney

As discussed above, baseline generation of LO products in the kidney is likely low. Under stimulated conditions, however, appreciable generation of

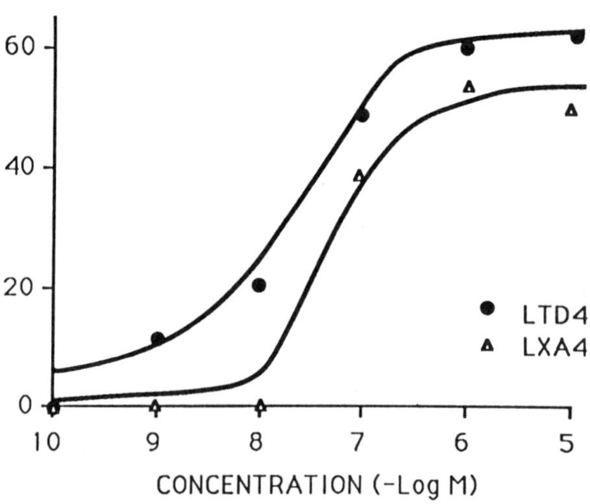

Fig. 5-3. Percent inhibition of ^{3}H-LTD_4 (10 nmol/L) binding to rat glomerular mesangial cells by LTD_4 and LXA_4. Each point represents the mean of four experiments performed in duplicate. (From Badr et al.,[45] with permission.)

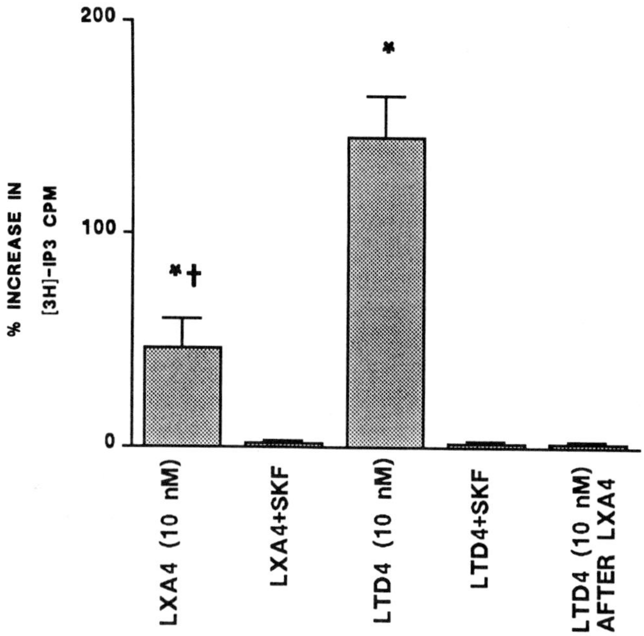

Fig. 5-4. $InsP_3$ formation in rat mesangial cells in response to LXA_4 and LTD_4 and its complete abolition in the presence of 100-fold concentration of SKF 104353, a specific LTD_4 receptor antagonist (SKF). The far right column depicts the abrogation of LTD_4-induced stimulation of $InsP_3$ formation by pre-exposure of the cells to 100 nmol/L LXA_4 for 10 minutes. *, $P < 0.01$ versus vehicle-stimulated controls; †, $P < .05$ versus LTD_4 alone. (From Badr et al.,[45] with permission.)

LTC_4, LTB_4, 12-HETE, 5-HETE, and 11-HETE have been documented. Early (24 hour) anti-glomerular basement membrane (nephrotoxic serum [NTS]) nephritis is characterized by increased glomerular synthesis of LTC_4,[30] LTB_4, 5-HETE, and 12-HETE,[26,28,29] and LTB_4 generation is also enhanced in cationic bovine γ-globulin-induced glomerulopathy.[27] The isolated perfused kidney from rejecting renal allografts[47] or antigen-stimulated preimmunized rats[21] releases increased amounts of LTC_4 and LTD_4 into the venous effluent, as does the acutely hydronephrotic kidney.[48]

At the cellular level, mesangial cells have been shown to generate 11- and 12-HETE following stimulation of phagocytosis by coincubation with serum-treated zymosan.[20] Since these cells are incapable of generating these products under baseline conditions, these in vitro experiments raise the intriguing possibility that LO pathway enzymes might be induced in vivo under the stimuli of inflammatory injury or macromolecular trafficking through the mesangium.

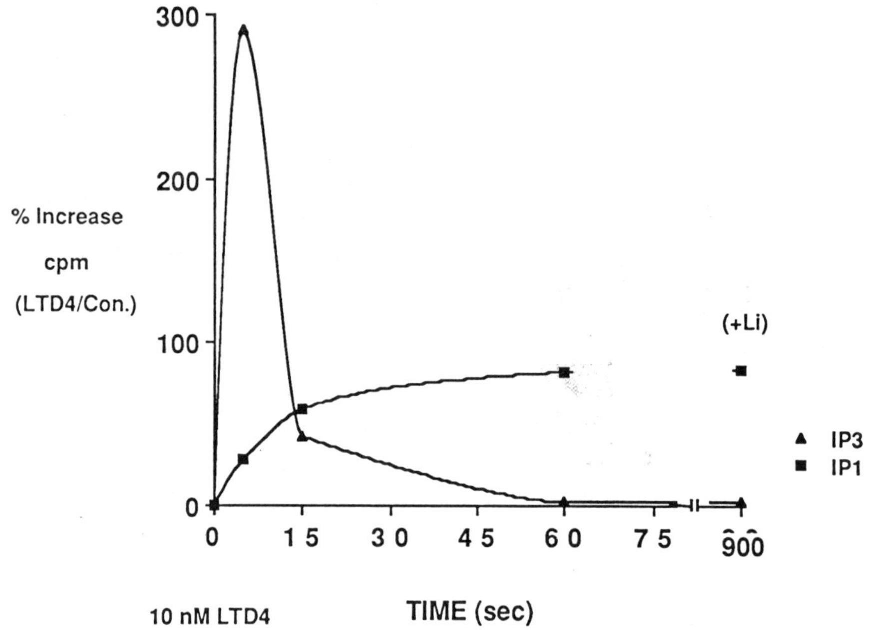

Fig. 5-5. Time course of stimulation of IP_1 and IP_3 following the addition of 10 nmol/L LTD_4 to rat glomerular mesangial cells. Mean radioactive counts obtained in vehicle-treated controls for IP_1 were 12,344, 8,971, 16,795, and 8,626 cpm at 5, 15, 60, and 900 seconds, respectively, and those for IP_3 were 4,011, 5,637, and 6,251 cpm at 5, 30, and 60 milliseconds, respectively. The rapid and transient rise in IP_3 generation at 5 seconds is followed by a gradual and time-dependent increase in IP_1 when the latter is measured in the presence of LiCl. The time course of appearance of ^3H-IP_2 (not shown) was variable and intermediate between the ^3H-IP_3 and ^3H-IP_1 peaks.

Biologic Activities of Lipoxygenase Products in the Kidney

Leukotrienes

Systemic administration of LTC_4 in the rat[2] and administration of LTC_4 and LTD_4 in the isolated perfused kidney[6] revealed potent renal vasoconstrictor actions of these eicosanoids. Micropuncture measurements revealed that LTD_4 exerts preferential constrictor effects on postglomerular arteriolar resistance and depresses the glomerular capillary ultrafiltration coefficient (K_f) and glomerular filtration rate (GFR).[3] The latter is likely due to receptor-mediated contraction of glomerular mesangial cells, which has been demonstrated for LTC_4 and LTD_4 in vitro.[43,44] These actions of LTD_4 in the kidney are consistent with its known smooth muscle contractile properties.[1] LTB_4, a potent chemotactic and leukocyte-activating agent,[1] is devoid of constrictor action in the normal rat kidney.[49] In addition to cellular contraction, these LTs also elicit mitogenic responses in rat and human mesangial cells and in glomerular epithelial cells.[37,43–46]

Lipoxins

LXA_4 dilates afferent arterioles when infused into the renal artery, without affecting efferent arteriolar tone.[45,50] This results in elevations in intraglomerular pressure and plasma flow rate, thereby augmenting GFR. The LXA_4-induced increase in GFR, however, is partially offset by its mild effect in decreasing K_f.[50] This latter effect is likely due to the aforementioned activation of LTD_4 binding sites on these cells by LXA_4.[45] Figure 5-6 summarizes LT/LX interactions in the glomerular microcirculation. It now appears likely that LXA_4-induced vasodilator responses in the afferent arteriole are mediated by the secondary release of cyclooxygenase products, such as prostaglandin E_2 (PGE_2).[51] LXA_4 has been demonstrated to release PGI_2 from cultured endothelial cells.[52] Other LXs (LXB_4 and 7-*cis*,11-*trans*-LXA_4) have been recently shown to possess vasoconstrictor actions in the rat kidney, the mechanisms of which are still undefined.[53]

Other Effects

Since it is likely that LO metabolites can be released near renal epithelia when there is surrounding inflammation, it is pertinent to ask whether these compounds can interfere with epithelial transport. In a recent study,[54] 5-HETE was shown to be taken up by cultured MDCK cells and incorporated into cell membrane phospholipid. The subsequent capacity of these cells to produce PGE_2 was reduced by more than 50 percent, suggesting a potential functional impact. Other relevant actions of LO products include inhibition by LTD_4 of atrial natriuretic factor-dependent cyclic guanosine monophosphate (cGMP) production in rat glomeruli, which may exaggerate its vasoconstrictive potential,[55] and involvement of 12-HETE in mediating angiotensin II feedback inhibition of renin release.[56]

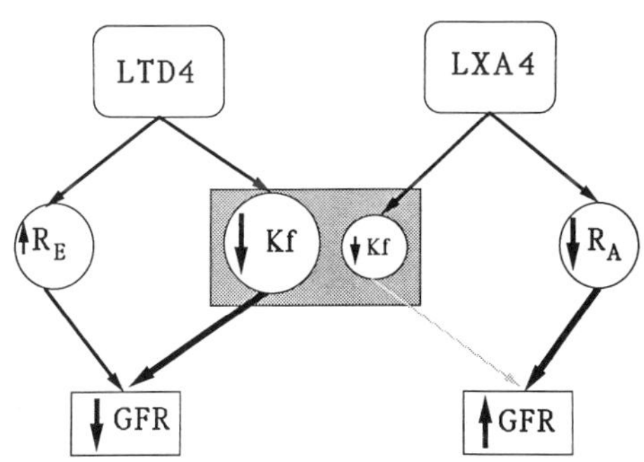

Fig. 5-6. Opposing effects of LTD_4 and LXA_4 on GFR and renal vascular resistance. Both eicosanoids decrease Kf through activation of a common receptor on glomerular mesangial cells (see text). R_E, efferent resistance; R_A, afferent resistance; Kf, glomerular capillary ultrafiltration coefficient.

Involvement of Lipoxygenase Products in Renal Pathophysiology

Studies of LO products in renal pathophysiology have been largely experimental due to the lack of availability of safe and effective LO inhibitors for human administration. The significance of the early burst of 5-LO activity during experimental glomerulonephritis has been highlighted by studies using a specific LTD_4 antagonist,[57] exogenous administration of LTB_4,[58] and dietary manipulations aimed at reducing LT synthetic capacity.[24,29,59] The data from these studies assign a major role for LTD_4 in mediating the fall in K_f and GFR that characterizes early NTS nephritis[57] and for LTB_4-mediated leukocyte recruitment and activation in amplifying the functional and histologic consequences of NTS administration.[58] Figure 5-7 summarizes these interactions. Lipoxygenase products appear to be involved in the recruitment, as well as the attachment,[60] of inflammatory cells to the glomerulus, thereby providing a potentially major stimulus for the perpetuation of injury. Interestingly, recent reports regarding inhibitory influences for LXs on neutrophil responses to chemotactic stimuli[61] and modulation of neutrophil-mesangial cell adherence[62] suggest a potential counter-regulatory influence for LXs on this aspect of the inflammatory response, thereby complementing their functional and receptor antagonism of LT actions.

CYTOCHROME P-450 METABOLITES OF ARACHIDONIC ACID

Chemistry and Nephronal Distribution

The presence of a microsomal NADPH-dependent cytochrome P-450 enzyme system in the mammalian kidney is now firmly established.[63–70] This pathway for the metabolism of AA (Fig. 5-8) has been identified in cortical and medullary tissues of pig, rabbit, and rat kidneys, and P-450/AA-derived metabolites have been demonstrated in human urine.[63–70] In fact, Schwartzman et al.[66] have shown that the specific activity for the formation of arachidonate metabolites via the P-450 pathway in the outer medulla of the rabbit kidney is even higher than that of the liver (7.9 μg of AA transformed/30 min/nmol of cytochrome P-450 for microsomes obtained from outer medulla versus 2.5 μg for liver-derived microsomes). The renal cytochrome P-450 system catalyzes the enzymatic transformation of AA via either an epoxygenase system yielding 5,6, 8,9, 11,12, and 14,15 epoxides (which can be hydrolyzed to their respective vicinal diols) or a monooxygenase system yielding w and w-1 oxidation products (Fig. 5-7). Complete characterization of this pathway in the rat kidney and demonstration of endogenous product generation by gas chromatography/mass spectrometry has been recently reported.[70] The major portion of renal enzymatic activity resides in the proximal tubule and thick ascending limb of Henle.[63,67]

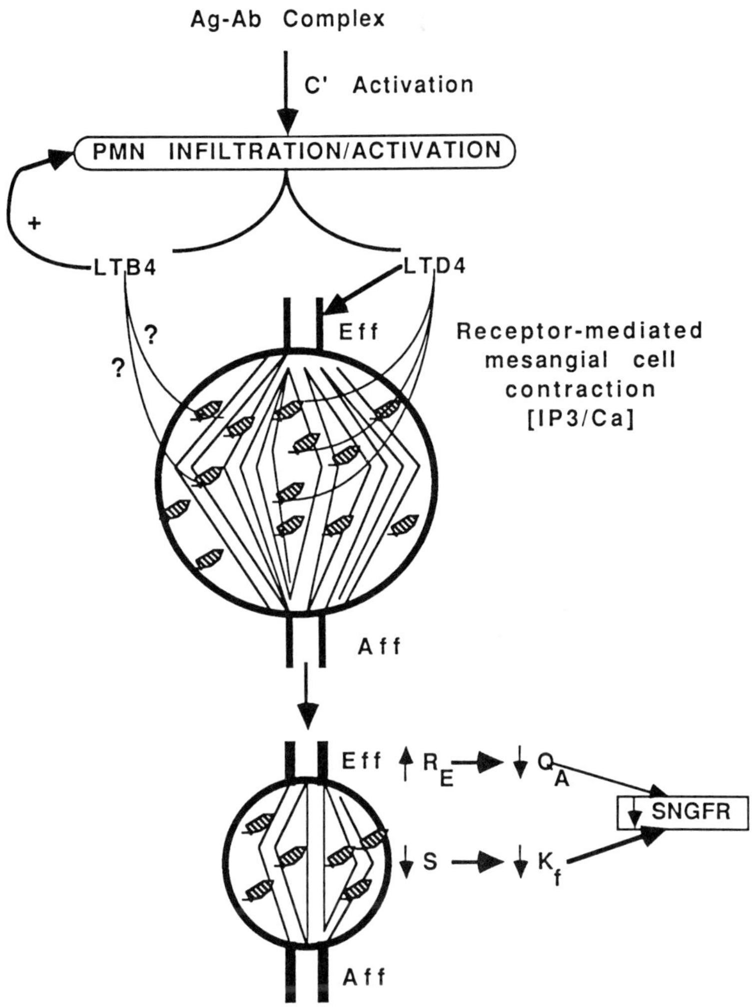

Fig. 5-7. Proposed mechanism for the role of LTs in the pathophysiology of immune-mediated glomerular injury. Antigen (Ag)–antibody (Ab) interactions within the glomerulus are associated with complement (C′) activation and polymorphonuclear (PMN) cell infiltration/activation. The latter results in the local release of LTB_4 and LTD_4. LTD_4 constricts efferent (Eff) arterioles and occupies specific receptors on the surface of smooth muscle-containing mesangial cells. Receptor activation in these cells is associated with the intracellular release of inositol triphosphate (IP3), an increase in intracellular Ca, and cell contraction. Mesangial cell contraction results in a reduction in the number of patent capillary loops, thereby decreasing glomerular size and filtering surface area (S), a component of the glomerular capillary ultrafiltration coefficient, Kf. The increase in efferent (R_E), as well as afferent (R_A), arteriolar resistances results in a fall in single nephron plasma flow rate (Q_A). The combined reductions in Q_A and Kf conspire to depress single nephron glomerular filtration rate (SNGFR). LTB_4 amplifies these responses through its chemoattractant and PMN-activator properties.

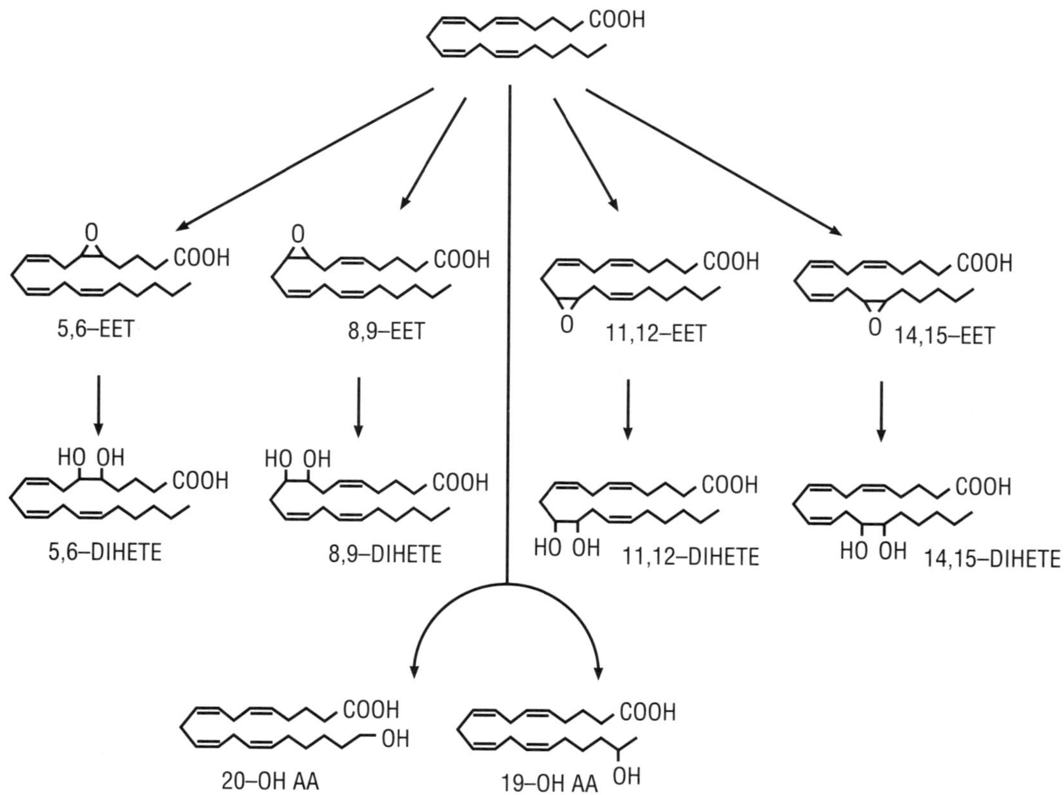

Fig. 5-8. Cytochrome P-450–mediated metabolism of AA.

Stimuli for Renal Release

A number of maneuvers have been shown to induce enzymatic activity catalyzing P-450–mediated oxygenation of AA in the kidney, including desoxycorticosterone acetate therapy following adrenalectomy,[71] unilateral nephrectomy,[72] experimental diabetes,[72] and pregnancy.[72] In addition, ischemic renal injury[73,74] and unclipping of the stenosed kidney in Goldblatt hypertensive rats[75] have been associated with the release of lipids, inhibitable with P-450 antagonists. The induction of this enzyme system in the aforementioned conditions may represent a physiologic adaptation aimed at increasing salt excretion and modulating renal hemodynamics (see below).

Renal Actions of P-450–Derived Metabolites of Arachidonic Acid

Certain of the epoxide derivatives of this pathway possess vasodilator or vasoconstrictor properties in various vascular beds.[76–80] In the kidney, 5,6-EET is a moderate vasoconstrictor whose action is likely mediated through the secondary generation of a cyclooxygenase product.[70] The vasoconstrictor action of this and other epoxides may limit the degree of vasorelaxation following unilateral nephrectomy, since inhibition of P-450–mediated AA metabolism 2 weeks postnephrectomy results in further vasodilatation of the remaining kidney and augmentation of GFR.[81] The involvement of this pathway in the pathogenesis of hypertension has been proposed, but sufficient evidence is still lacking.[82–85]

Effects on Transport

Proximal tubule. The proximal tubule is a major source for cytochrome P-450 enzyme in the kidney.[63,86] It is likely that most studies characterizing renal cortical production of P-450 metabolites of AA are really characterizing proximal tubule production. Although some qualitative difference between rat and rabbit kidney cortex is observed, microsomal fractions of renal cortex demonstrate that the majority of arachidonate metabolites are produced via the cytochrome P-450 enzyme system. In the rabbit kidney cortex, the predominant metabolites are omega and omega minus one products, with EETs and DHETs being secondary; in the rat cortex, the EETs and DHETs are produced in greater quantities than the omega and omega minus one oxidation products.[70,87] Studies of isolated microdissected proximal straight tubules demonstrate the production of both 11,12- and 14,15-EET and their corresponding diols. Interestingly, there was little evidence for omega and omega minus one oxidation products, suggesting that the formation of these latter compounds may occur in other nephron segments or are more selectively made with microsomal preparations than with intact cells.[87]

If one considers that the cytochrome P-450 metabolites of AA, like the cyclooxygenase metabolites, likely function as autacoids, then it is logical to search for some functional effect of these compounds in the proximal tubule. No studies have been performed examining salt and water transport, but preliminary studies have examined for a relationship between P-450 metabolism of AA and exposure of proximal straight tubules to the organic anion para-aminohippuric acid (PAH). Proximal straight tubules secrete PAH, but do not metabolize it through any mechanism involving cytochrome P-450. When tubules are exposed to PAH, there is a dramatic increase in P-450 metabolism of AA[88] (Fig. 5-9). It is not clear whether these observations can be interpreted to provide a causal link between organic anion secretion and P-450 metabolism of AA.

To search for a possible link between P-450 metabolism of AA and overall renal handling of salt, cortical and medullary microsomal metabolism of AA through cytochrome P-450 was measured in rabbits given daily injections of desoxycorticosterone acetate, 5 mg/kg body weight, for 21 days.[71] Both cortical and medullary metabolism of AA via cytochrome P-450 was significantly increased with desoxycorticosterone acetate treatment, with peak response appearing at day 5 and being maintained until withdrawal. Although the increase in metabolism as a percentage change was greater in medullary microsomes, the absolute increase in cortex was dramatic.[71] It is likely, but not definite, that the cortical metabolism represents proximal tubule metab-

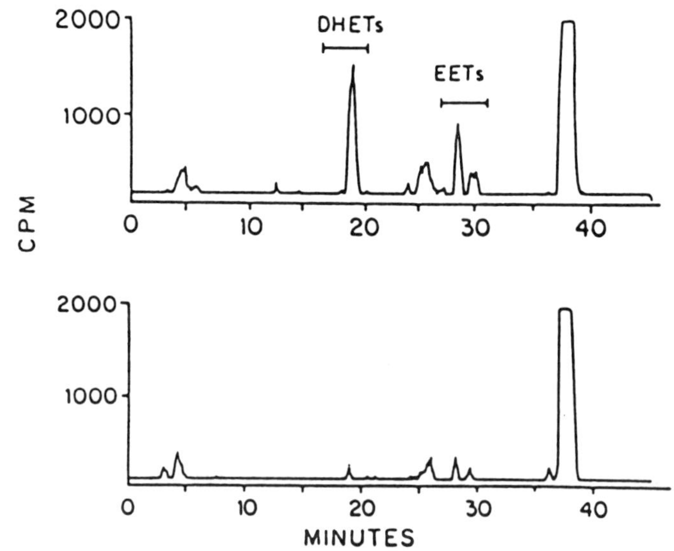

Fig. 5-9. Effect of PAH on the oxidative metabolism of AA by isolated rabbit kidney pars recta. Pars recta (40 mm) were incubated 30 minutes at 37°C with $^{3}H_{8}$-AA (1 μmol/L, 1 Ci/mmol) in the presence (top) or absence (bottom) of PAH. Details of product extraction and resolution are given in the text. High-performance liquid chromatography column eluents were monitored for ^{3}H with an on-line Bis β detector. The chromatograms obtained by injecting equivalent aliquots of the organic soluble products extracted from one paired experiment are shown.

olism. In any event, these studies suggest that the cytochrome P-450 that metabolizes AA may be ineducable by desoxycorticosterone acetate.

Thick ascending limb of henle. Schwartzman and colleagues have demonstrated that isolated thick ascending limb cells convert AA to biologically active metabolites via cytochrome P-450,[68] one of which inhibits purified sodium potassium ATPase.[68,85] Interestingly, the formation of these oxygenated arachidonate metabolites by thick ascending limb of Henle cells is stimulated by two peptide hormones, arginine vasopressin and salmon calcitonin.[89] More recently, these investigators have shown that there is an increased production of these compounds in thick ascending limb of Henle cells obtained from rabbits made hypertensive by aortic constriction.[76] These investigators postulate that one of the compounds may be an endogenous inhibitor of sodium potassium ATPase and may, under certain circumstances, exhibit a protective effect to limit the degree of thick ascending limb of Henle cell injury in response to hypoperfusion by decreasing energy demands of sodium potassium ATPase. This interesting hypothesis requires more support.

A recent observation demonstrates an increase in the P-450–dependent metabolism of AA in the kidneys of young spontaneously hypertensive rats while their blood pressure is rapidly increasing. This increased P-450 metabolism was not observed in the kidneys of adult spontaneously hypertensive rats or in another strain of rats, the Wistar/Kyoto rats, with normal blood pressure.[83] In an attempt to causally link the elevated blood pressure with increased P-450 metabolism of arachidonate, the investigators treated both spontaneously hypertensive rats and Wistar/Kyoto rats with stannous chloride, which selectively depletes renal cytochrome P-450. This treatment restored blood pressure to normal in the spontaneously hypertensive rats and also produced a natriuresis.[85] There was also an associated decreased generation of P-450 arachidonate metabolite in the kidneys of the young spontaneously hypertensive rats. This provocative observation raises a number of questions that need to be addressed, including the cause and effect relationship between inhibition of P-450 arachidonate metabolism and the observed reduction of blood pressure and increase in sodium excretion. Also, the specific product or products responsible and their source within the kidney need to be identified.

Collecting duct. Early studies with biosynthetically produced EETs examined their effect on the in vitro perfused rabbit cortical collecting duct.[71] These studies demonstrated that the 5,6-EET, when placed in the perfusion solution, inhibited sodium absorption and lumen-negative transepithelial voltage. The mechanism responsible for the 5,6-EET inhibition of sodium transport, and whether this compound is active from the bath side of the tubule remain to be explored.

Initial studies on toad urinary bladders have shown that several of the EETs and DHETs at high concentrations inhibit vasopressin-induced osmotic water flow.[90] This important early observation has led to subsequent studies of the mammalian cortical collecting duct perfused in vitro. The effects of all four DHETs on arginine vasopressin-induced water flow in the rabbit cortical collecting duct were examined. While all four DHETs inhibited osmotic water flow, the 14,15-DHET was most potent. Indeed, in a paired-dose response study, the 14,15-DHET was as potent as PGE_2 in inhibiting AVP-induced osmotic water flow.[91] In contrast to the effects of PGE_2, however, the 14,15-DHET had little effect on transepithelial voltage in the cortical collecting duct, suggesting that it has no effect on sodium transport. Also, in contrast to the previously published studies of PGE_2, the 14,15-DHET inhibited cyclic adenosine monophosphate-induced osmotic water flow in the cortical collecting duct. Thus, the 14,15-DHET has a post-cyclic adenosine monophosphate component to its inhibition of water flow. To provide physiologic relevance to studies with the isolated perfused tubule, production of P-450 metabolites of AA was analyzed in cortical collecting duct cells harvested with the use of monoclonal antibodies. With the use of gas chromatography and mass spectrometry, these studies clearly demonstrated that 14,15-EET and DHET are produced by cortical collecting duct.[91] It will be important to determine the regulation of P-450 metabolism of arachidonate in cortical collecting duct cells.

In summary, the cytochrome P-450 metabolites of AA are the most recently described family of potentially important biologically active metabolites of this fatty acid. The kidney appears to be a rich source for these metabolites. It is likely that they have some important biologic actions within the kidney. It is relevant, therefore, to note that these compounds have been identified in the normal human urine using gas chromatographic/ mass spectrometric techniques. Of note, their concentration in the urine is equivalent to that seen with cyclooxygenase metabolites.[69]

PLATELET-ACTIVATING FACTOR

Platelet-activating factor (PAF) is a family of 1-alkyl-2-acetyl glycero-3-phosphocholines produced by inflammatory cells (monocyte/macrophage, polymorphonuclear neutrophil, eosinophil, basophil, mast cell), platelets,[92,93] endothelial cells,[94] isolated glomeruli, mesangial cells,[95–98] and renal medullary cells.[92] Its intracellular generation is stimulated by a number of experimental and pathophysiologic stimuli, such as endotoxin,[99] complement C5a[29] and C3a,[96] leukotrienes,[94] platelet-derived growth factor,[96] calcium ionophore,[100] vasoactive peptides, formyl-methionyl-leucyl-phenyl-alanine, and others,[92] as well as during phagocytosis.[101] Due to the concomitant cytosolic release of AA that accompanies PAF biosynthesis,[92] its formation has been strongly linked to the generation of the biologically active cyclooxygenase and LO products of AA, and the interactions among these products and PAF represent a crucial component of the expression of its biologic effects. The latter include platelet aggregation and activation,[102] chemotaxis and chemokinesis,[103] polymorphonuclear leukocyte aggregation and degranulation,[104] stimulation of oxygen-free radical formation,[105] smooth muscle contraction,[92,106,107] and increased vascular permeability.[92,107]

Activated leukocytes of macrophage/monocyte origin, as well as platelets, are known to infiltrate and/or proliferate within the glomerulus during inflammatory reactions. Hence, under these conditions, release of PAF into the glomerular microenvironment is expected. Evidence exists, however, to suggest that PAF is also released from normal glomeruli or from kidneys subjected to noninflammatory forms of injury, such as endotoxin,[21,99] calcium ionophore,[98,101] and antigen challenge.[21] The presence, within the normal glomerular mesangium, of Ia-positive, bone marrow-derived, phagocytic cells characterized as monocytes raises the strong possibility that in the absence of an overt inflammatory infiltrate, intraglomerular PAF production may occur through the stimulation and proliferation of these cells. More recent studies by Schlondorff and colleagues[98] and by Lianos and Zanglis[96] have confirmed that cultured mesangial cells of the smooth muscle containing subpopulation can produce PAF when stimulated with calcium ionophore, C3a, platelet-derived growth factor, and PAF itself. This may assign a major role for glomerular mesangial cells as a source for PAF during inflammatory and non-inflammatory forms of glomerular injury. Finally, glomerular endothelial cells could represent a source of PAF, as has been demonstrated in other tissues.[94] In view of its role as a proinflammatory mediator in glomerular injury, we investigated the renal cortical microcirculatory responses to the intrarenal arterial administration of PAF in the anesthetized euvolemic Munich/Wistar rat.[108] Closed arterial administration of PAF led to dose-dependent reductions in renal plasma flow rate, GFR, and filtration fraction in the absence of hypotension or hemoconcentration. PAF increased pre- and postglomerular arteriolar resistances and led to a dramatic reduction in the mean value for Kf. PAF-induced changes in renal hemodynamics were abolished in the presence of the cyclooxygenase inhibitors, indomethacin and ibuprofen. When administered concomitantly with a thromboxane A_2 receptor antagonist, PAF led to significant increases in renal plasma flow and GFR. In isolated glomeruli, PAF stimulated the biosynthesis of thromboxane B_2 in a dose-dependent manner. Thus, PAF depressed rat glomerular function by inducing contraction of arteriolar and mesangial smooth muscle, likely via the secondary release of thromboxane A_2.

ACKNOWLEDGMENT

Studies from the author's laboratory cited in this chapter were supported by the Vanderbilt Kidney and Urologic Disease Center Grant (DK 39261) and NIH Grants DK 38667 and DK 38226.

REFERENCES

1. Samuelsson B, Dahlen S-E, Lindgren JA et al: Leukotrienes and lipoxins: structures, biosynthesis, and biological effects. Science 237:1171, 1987
2. Badr KF, Baylis C, Pfeffer JM et al: Renal and systemic hemodynamic responses to intravenous infusion of leukotriene C4 in the rat. Circ Res 54:492, 1984
3. Badr KF, Brenner BM, Ichikawa I: Effects of leukotriene D4 on glomerular dynamics in the rat. Am J Physiol 253:F239, 1987
4. Simonson MS, Dunn MJ: Leukotriene C4 and D4 contract rat glomerular mesangial cells. Kidney Int 30:524, 1986
5. Barnett R, Goldwasser P, Scharschmidt LA, Schlondorff D: Effects of leukotrienes on isolated rat glomeruli and cultured mesangial cells. Am J Physiol 250:F838, 1986
6. Rosenthal A, Pace-Asciak CR: Potent vasoconstriction of the isolated perfused kidney by leukotrienes C4 and D4. Can J Pharmacol 61:325, 1983
7. Matsumoto T, Funk CD, Radmark O et al: Molecular cloning and amino acid sequence of human 5-lipoxygenase. Proc Natl Acad Sci USA 85:26, 1988
8. Funk CD, Radmark O, Fu JY et al: Molecular cloning and amino acid sequence of leukotriene A4 hydrolase. Proc Natl Acad Sci USA 84:6677, 1987
9. Fitzpatrick F, Ligget W, McGee J et al: Metabolism of leukotriene A4 by human erythrocytes. J Biol Chem 259:11403, 1984
10. Feinmark SJ, Cannon PJ: Endothelial cell leukotriene C4 synthesis results from intercellular transfer of leukotriene A4 synthesized by polymorphonuclear leukocytes. J Biol Chem 261:16466, 1986
11. Marcus A, Broekman M, Safier ML et al: Formation of leukotrienes and other hydroxy acids during platelet-neutrophil interactions in vitro. Biochem Biophys Res Commun 109:130, 1982
12. Maclouf JA, Murphy RC: Transcellular metabolism of neutrophil-derived leukotrien A4 by human platelets. A potential cellular source of leukotriene C4. J Biol Chem 263:174, 1988
13. Nicolau KC, Marron BE, Veale CA et al: Identification of a vowel 7-cis-11-trans-lipxin A4 generated by human neutrophils total synthesis, spasmogenic activities, and comparison with other geometric isomers of lipoxin A4 and B4. Biochim Biophys Acta 1003:44, 1989
14. Serhan CN, Sheppard K-A: Lipoxin formation during human neutrophil-platelet interactions. Evidence for the transformation of leukotriene A4 by platelet 12-lipoxygenase *in vitro.* J Clin Invest 85:772, 1990
15. Livio M, Chiabrando C, Macconi D et al: Metabolism of arachidonic acid in isolated glomeruli from pig kidney. Biochim Biophys Acta 961:110, 1988
16. Cattell V, Cook HT, Smith J et al: Leukotriene B4 production in normal rat glomeruli. Nephrol Dial Transplant 2:154, 1987
17. Sraer J, Rigaud M, Bens M et al: Metabolism of arachidonic acid via the lipoxygenase pathway in human and murine glomeruli. J Biol Chem 258:4325, 1983
18. Ardaillou R, Baud L, Sraer J: Leukotrienes and other lipoxygenase products of arachidonic acid synthesized in the kidney. Am J Med, suppl. 2B 81:12, 1986
19. Jim K, Hassid A, Sun F, Dunn MJ: Lipoxygenase activity in rat kidney glomeruli, glomerular epithelial cells, cortical tubules. J Biol Chem 257:10294, 1982
20. Baud L, Hagege J, Sraer J et al: Reactive oxygen production by cultured rat

glomerular mesangial cell during phagocytosis is associated with stimulation of lipoxygenase activity. J Exp Med 158:1836, 1983
21. Pirotzky E, Bidault J, Burtin C et al: Release of platelet activating factor, slow reacting substance and vasoactive amines from isolated rat kidney. Kidney Int 25:404, 1984
22. Garrick R, Shen S-Y, Ogunc S, Wong PY-K: Transformation of leukotriene A4 to lipoxins by rat kidney mesangial cell. Biochem Biophys Res Commun 162:626, 1989
23. Sraer J, Baud L, Bens M et al: Glomeruli cooperate with macrophages in converting arachidonic acid to prostaglandins and hydroxyeicosatrienoic acids. Prostaglandins Leukot Med 13:67, 1984
24. Lefkowith JB, Morrison AR, Schreiner GF: Murine glomerular leukotriene B4 synthesis. Manipulation by (n-6) fatty acid deprivation and cellular origin. J Clin Invest 82:1655, 1988
25. Lefkowith JB, Schreiner G: Essential fatty acid deficiency depletes rat glomeruli of resident macrophages and inhibits angiotensin II-induced eicosanoid synthesis. J Clin Invest 80:947, 1987
26. Lianos AA: Synthesis of hydroxyeicosatetraenoic acids and leukotrienes in rat nephrotoxic serum glomerulonephritis. J Clin Invest 82:427, 1988
27. Rahman MA, Nakazawa M, Emancipator SN, Dunn MJ: Increased leukotriene B4 synthesis in immune injured rat glomeruli. J Clin Invest 81:1945, 1988
28. Fauler J, Wiemeyer A, Marx K-H et al: LTB4 in nephrotoxic serum nephritis in rats. Kidney Int 36:46, 1989
29. Schreiner GF, Rovin B, Lefkowith JB: The anti-inflammatory effects of essential fatty acid deficiency in experimental glomerulonephritis: the modulation of macrophage migration and eicosanoid metabolism. J Immunol 143:3192, 1989
30. Mune S, Goto T, Matoba K et al: Glomerular leukotriene synthesis in rat nephrotoxic serum nephritis. p. 232a. Proc Xth Int Congress Nephrol London, July 26-31, 1987
31. Winokur TS, Morrison AR: Regional synthesis of monohydroxy eicosanoids by the kidney. J Biol Chem 256:10221, 1981
32. Sraer J, Bens M, Oudinet JP, Ardaillou R: Bioconversion of leukotriene C4 by rat glomeruli and papilla. Prostaglandins 31:909, 1986
33. Hirota T, Nishikawa Y, Takahagi G et al: Simultaneous purification and properties of dehydropeptidase-I and aminopeptidase-M from rat kidney. Res Commun Chem Pathol Pharmacol 49:435, 1985
34. Breuer O, Hammarstrom S: Enzymatic conversion of leukotriene B4 to 6-trans-leukotriene B4 by rat kidney homogenates. Biochem Biophys Res Commun 142:667, 1987
35. Kaever V, Martin M, Fauler J et al: A novel metabolic pathway for leukotriene b4 in different cell types: primary reduction of a double bond. Biochim Biophys Acta 922:337, 1987
36. Ballermann BJ, Lewis RA, Corey EJ et al: Identification and characterization of leukotriene C4 receptors isolated rat glomeruli. Circ Res 56:324, 1985
37. Baud L, Sraer J, Perez J et al: Leukotriene C4 binds to human glomerular epithelial cells and promotes their proliferation in vitro. J Clin Invest 76:374, 1985
38. Sun FF, Chau LY, Austen KF: Binding of leukotriene C4 by glutathione trans-

ferase: a reassessment of biochemical and functional criteria for leukotriene receptors. Fed Proc 46:204, 1987

39. Mong S, Hoffman K, Wu H-L, Crooke ST: Leukotriene-induced hydrolysis of inositol lipids in guinea pig lung: mechanism of signal transduction for leukotriene-D4 receptors. Mol Pharmacol 31:35, 1987
40. Mong S, Miller J, Wu H-L, Crooke ST: Leukotriene D_4 receptor-mediated hydrolysis of phosphoinositides and mobilization of calcium in sheep tracheal smooth muscle cells. J Pharmacol Exp Ther 244:508, 1988
41. Mong S, Wu HL, Miller J et al: SKF 104353, a high affinity antagonist for human and guinea pig lung leukotriene D_4 receptor blocked phosphoinositide metabolism and thromboxane synthesis induced by LTD_4. Mol Pharmacol 32:223, 1987
42. Sarau HM, Mong S, Foley S et al: Identification and characterization of leukotriene D_4 receptors and signal transduction processes in rat basophilic leukemia cells. J Biol Chem 262:4034, 1987
43. Badr KF, Mong S, Hoover RL et al: Leukotriene D4 binding and signal transduction in rat glomerular mesangial cells. Am J Physiol 257:F280, 1989
44. Simonson MS, Mene P, Dubyak GR, Dunn MJ: Identification and transmembrane signaling of leukotriene D4 receptors in human mesangial cells. Am J Physiol 255:C771, 1988
45. Badr KF, DeBoer DK, Schwartzberg M, Serhan CN: Lipoxin A4 antagonizes cellular and in vivo actions of leukotriene D4 in rat glomerular mesangial cells: evidence for competition at a common receptor. Proc Natl Acad Sci USA 86:3438, 1989
46. Baud L, Perez J, Cherqui G et al: Leukotriene D4-induced proliferation of glomerular epithelial cells: PKC and Na+-H+ exchanger-mediated response. Am J Physiol 257:C232, 1989
47. Coffman TM, Yarger WE, Klotman PE: Leukotriene C4 production is increased in rejecting rat renal allografts. Kidney Int 29:332, 1986
48. Albrightson CR, Evers AS, Griffin AC, Needleman P: Effect of endogenously produced leukotrienes and thromboxane on renal vascular resistance in rabbit hydronephrosis. Circ Res 61:514, 1987
49. Yared A, Takahashi K, Brash A et al: Vasorelaxant and natriuretic responses to leukotriene B4 in the rat kidney: functional importance of the 12(R) stereochemistry. Kidney Int 35:298, 1989
50. Badr KF, Serhan CN, Nicolau KC, Samuelsson B: The action of lipoxin A4 on glomerular microcirculatory dynamics in the rat. Biochem Biophys Res Commun 145:408, 1987
51. DeBoer DK, Ebert J, Schwartzberg M et al: Lipoxin A4-induced renal vasodilatation is cyclooxygenase-dependent. Kidney Int 37:347, 1990
52. Brezinski ME, Gimbrone MA, Jr, Nicolau KC, Serhan CN: LIpoxins stimulate prostacycli generation by human endothelial cells. FEBS Lett 245:167, 1989
53. Takahashi K, DeBoer DK, Serhan CN, Badr KF: Vasoconstrictor lipoxins: renal vascular actions of lipoxin B4 and of a novel eicosanoid: 7-cis,11-trans-lipoxin A4. Kidney Int 37:351, 1990
54. Gordon JA, Figard PH, Quinby GE, Spector AA: 5-HETE: uptake, distribution, and metabolism in MDCK cells. Am J Physiol 256:C1, 1989
55. Terada Y, Tomita K, Yoshiyama M et al: Leukotriene D4 inhibits atrial natriuretic factor-dependent cGMP production in rat glomerulus. Am J Physiol 256:F95, 1989

56. Antonipillai L, Nadler J, Horton R: Angiotensin feedback inhibition on renin is expressed via the lipoxygenase pathway. Endocrinology 122:1277, 1988
57. Badr KF, Schreiner GF, Wasserman MA, Ichikawa I: Preservation of the glomerular capillary ultrafiltration coefficient during rat nephrotoxic serum nephritis by a specific leukotriene D4 receptor antagonist. J Clin Invest 81:1702, 1988
58. Badr KF: Leukotriene D4/leukotriene B4 interactions in the pathophysiology of experimental glomerulonephritis. Adv Prostaglandin Thromboxane Leukotriene Res 19:233, 1989
59. Takahashi K, Schreiner GF, Lefkowith JB, Badr KF: Essential fatty acid deficiency normalizes glomerular functions and suppresses leukotriene/thromboxane synthesis in nephrotoxic serum nephritis. Kidney Int 37:352, 1990
60. Baud L, Sraer J, Delarue F et al: Lipoxygenase products mediate the attachment of rat macrophages to glomeruli in vitro. Kidney Int 27:855, 1985
61. Lee TH, Horton CE, Kyan-Aung U et al: Lipoxin A4 and lipoxin B4 inhibit chemotactic responses of human neutrophils stimulated by leukotriene B4 and N-formyl-L-methionyl-L-leucyl-L-phenylalanine. Clin Sci 77:195, 1989
62. Brady HR, Persson U, Brenner BM, Serhan CN: Leukotrienes and lipoxins modulate neutrophil-mesangial cell adhesion: role of CD18/CD11 complex. Clin Res 38:275A, 1990
63. Endou H: Cytochrome P-450 monooxygenase system in the rabbit kidney: its intranephron localization and its induction. Jpn J Pharmacol 33:423, 1983
64. Oliw EH, Lawson JA, Brash AR, Oates JA: Arachidonic acid metabolism in rabbit renal cortex. J Biol Chem 256:9924, 1981
65. Carroll M, Schwartzman M, Abraham N et al: Cytochrome P450-dependent arachidonate metabolism in renomedullary cells: formation of $Na^{+}K^{+}$-ATPase inhibitor. J Hypertens, suppl 4:S33, 1986
66. Schwartzman M, Abraham N, Carroll M et al: Regulation of arachidonic acid metabolism by cytochrome P-450 in rabbit kidney. Biochem J 238:283, 1986
67. Morrison A, Pascoe N: Metabolism of arachidonate through NADPH-dependent oxygenase of renal cortex. Proc Natl Acad Sci USA 78:7375, 1981
68. Schwartzman M, Ferreri N, Carroll M et al: Renal cytochrome P450-related arachidonate metabolite inhibits $(Na^{+} + K^{+})$ ATPase. Nature 314:620, 1985
69. Toto R, Siddhanta A, Manna S et al: Arachidonic acid epoxygenase: detection of epoxyeicosatrienoic acids (EETs) in human urine. Biochim Biophys Acta 919:132, 1987
70. Takahashi K, Capdevila J, Karara A et al: Cytochrome P450 arachidonate metabolites in rat kidney I: characterization and hemodynamic responses. Am J Physiol 258:F781, 1990
71. Jacobson HR, Corona S, Capdevila J et al: Effect of epoxyeicosatrienoic acids on ion transport in the rabbit collecting tubule. p. 311. In Braquet P, Garay RP, Frölich JC et al (eds): Prostaglandins and Membrane Ion Transport. Raven Press, New York, 1984
72. Badr KF, Takahashi K, Falck JR et al: Induction of cytochrome P450-linked metabolism of arachidonic acid following unilateral nephrectomy and during experimental diabetes and mid-term pregnancy in the mammalian kidney. Clin Res 36:513A, 1988
73. Brezis N, Rosen S, Epstein FH: Acute renal failure. p. 735. In Brenner BM, Rector FC, Jr (eds): The Kidney. 3rd Ed. WB Saunders, Philadelphia, 1986

74. Cantley L, Fuhro R, Silva P et al: Medullary thick ascending limb cells produce a transport inhibitor, abstracted. Kidney Int 29:351, 1986
75. Muiread EE, Byers LW, Capdevila J et al: Effect of cytochrome P-450 induction and inhibition on the antihypertensive function of the kidney. Trans Assoc Am Physicians 100:196, 1987
76. Carroll MA, Schwartzman M, Baba M et al: Renal cytochrome P-450-related arachidonate metabolism in rabbit aortic coarctation. Am J Physiol 24:F151, 1988
77. Carroll MA, Schwartzman M, Capdevila J et al: Vascular activity of arachidonic acid epoxides. Eur J Pharmacol 138:281, 1987
78. Coceani F, Breen CA, Lees JG et al: Further evidence implicating a cytochrome P-450-mediated reaction in the contractile tension of the lamb ductus arteriosus. Circ Res 62:471, 1988
79. Escalante B, Sessa WC, Falck JR et al: Vasoactivity of 20-hydroxyeicosatetraenoic acid is dependent on metabolism by cyclooxygenase. J Pharmacol Exp Ther 248:229, 1989
80. Proctor KG, Falck JR, Capdevila J: Intestinal vasodilation by epoxyeicosatrienoic acids: arachidonic acid metabolites produced by a cytochrome P450 monoxygenase. Circ Res 60:50, 1987
81. Takahashi K, Capdevila J, Munger K et al: A role for cytochrome P450 arachidonate metabolism in the contralateral renal hemodynamic response to uninephrectomy. Clin Res 37:503A, 1989
82. Escalante B, Falck JR, Yadagiri P et al: 19(S)-Hydroxyeicosatetraenoic acid is a potent stimulator of renal Na^+-K^+-ATPase. Biochem Biophys Res Commun 152:1269, 1988
83. Sacerdoti D, Abraham NG, McGiff JC, Schwartzman ML: Renal cytochrome P-450-dependent metabolism of arachidonic acid in spontaneously hypertensive rats. Biochem Pharmacol 37:521, 1988
84. Ferreri NF, Schwartzman M, Ibraham NG et al: Arachidonic acid metabolism in a cell suspension isolated from rabbit renal outer medulla. J Pharmacol Exp Ther 231:441, 1984
85. Sacerdoti D, Escalante B, Abraham NG et al: Treatment with tin prevents the development of hypertension in spontaneously hypertensive rats. Science 243:388, 1989
86. Dees J, Master BSS, Muller-Eberhard U, Johnson E: Effect of 2,3,7,8-tetrachlorodibenzo-p-dioxin and phenobarbital on the occurrence and distribution of four cytochrome P-450 isozymes in rabbit kidney, lung, and liver. Cancer Res 42:1423, 1982
87. Falck JR, Schueler VJ, Jacobson HR et al: Arachidonate epoxygenase: identification of epoxyeicosatrienoic acids in rabbit kidney. J Lipid Res 28:840, 1987
88. Lapuerta L, Chacos N, Falck JR et al: Renal microsomal cytochrome P-450 and the oxidative metabolism of arachidonic acid. Am J Med Sci 31:275, 1988
89. McGiff JC, Laniado-Schwartzman M: Arachidonic acid metabolites and blood pressure control. Clin Physiol Biochem 6:179, 1988
90. Schlonforff D, Petty E, Oates JA et al: Epoxygenase metabolites of arachidonic acid inhibit vasopressin response in toad bladder. Am J Physiol 253:F464, 1987
91. Hirt DL, Capdevila J, Falck JR et al: Cytochrome P450 metabolites of arachidonic acid are potent inhibitors of vasopressin action on rabbit cortical collecting duct. J Clin Invest 3:235, 1989

92. Schlondorff D, Neuwirth R: Platelet activating factor and the kidney. Am J Physiol 251:F1, 1986
93. Vargaftig BB, Chignard M, Benveniste J, Wal F: Background and present status of research on platelet activating factor (PAF-acether). Ann N Y Acad Sci 370:119, 1981
94. McIntyre TM, Zimmerman GA, Prescott S: Leukotrienes C4 and D4 stimulate human endothelial cells to synthesize platelet activating factor and bind neutrophils. Proc Natl Acad Sci USA 83:2204, 1986
95. Lefkowith JB, Schreiner G: Essential fatty acid deficiency depletes rat glomeruli of resident macrophages and inhibits angiotensin II-induced eicosanoid synthesis. J Clin Invest 80:947, 1987
96. Lianos EA, Zanglis A: Mesangial cell synthesis of platelet activating factor in response to defined inflammatory mediators. Kidney Int 33:273A, 1988
97. Schlondorff D: The glomerular mesangial cell: an expanding role for a specialized pericyte. FASEB J 1:272, 1987
98. Schlondorff D, Goldwasser P, Neuwirth R et al: Production of platelet activating factor in isolated glomeruli and cultured glomerular mesangial cells. Am J Physiol 250:F1123, 1986
99. Wang J, Dunn MJ: Platelet-activating factor mediates endotoxin-induced acute renal insufficiency in rats. Am J Physiol 253:F1283, 1987
100. Schwertslag U, Scherf H, Gerber JG et al: L-platelet activating factor (L-PAF) induced changes in renal vascular resistance, vascular reactivity, and renin release in the isolated perfused rat kidney, abstracted. Kidney Int 29:388A, 1986
101. Alonso FM, Garcia GL, Sanchez-Crespo M, Mato JM: Activation of 1-alkyl-2-lyso-glycero-3-phosphocholine:acetyl-CoA-transferase during phagocytosis in human polymorphonuclear leukocytes. J Biol Chem 257:3376, 1982
102. Shaw JO, Klusick SJ, Hanahan DJ: Activation of rabbit platelet phospholipase and thromboxane synthesis by 1-o-hexadecyl/octadecyl-2-acetyl-sn-glyceryl-3-phosphorylcholine (platelet activating factor). Biochim Biophys Acta 663:222, 1981
103. Wardlaw AJ, Moqbel R, Cromwell O, Kay AB: Platelet-activating factor. A potent chemotactic and chemokinetic factor for human eosinophils. J Clin Invest 78:1701, 1986
104. Shaw JO, Pinckard RN, Ferrigini KS et al: Activation of human neutrophils with 1-o-hexadecyl/octadecyl-2-acetyl-sn-glyceryl-3-phosphorylcholine (platelet activating factor). J Immunol 127:1250, 1981
105. Sedor JR, Abboud HE: Platelet activating factor stimulates oxygen radical release by cultured rat mesangial cells, abstracted. Kidney Int 27:222, 1985
106. Kester M, Mene P, Dubyak G, Dunn MJ: Elevation of cytosolic free calcium by platelet activating factor in cultured rat mesangial cells. FASEB J 1:219, 1987
107. Pirotzky E, Page C, Morley J et al: Vascular permeability induced by PAF-acether (platelet activating factor). Agents Actions 16:1, 1985
108. Badr KF, DeBoer DK, Takahashi K et al: Glomerular responses to platelet activating factor in the rat: role of thromboxane A2. Am J Physiol 256:F35, 1989

6

The Role of Cholesterol in Glomerular Injury

Jonathan R. Diamond

INTRODUCTION

Recently, much attention has been directed toward the secondary hyperlipidemia of nephrosis as a putative aggravating factor in the progression of initial glomerular injury to glomerulosclerosis. This chapter will review the experimental data that support the contention that hypercholesterolemia is capable of exacerbating glomerular injury. More recent information that identifies the glomerular macrophage as a potential mechanism for the aggravating effect of hypercholesterolemia on glomerular injury will be presented.

PATHOPHYSIOLOGIC MECHANISMS MEDIATING THE HYPERLIPIDEMIA OF NEPHROSIS

Proteinuria, hypoalbuminemia, edema, and hyperlipidemia are the four major features of the nephrotic syndrome. The development of this hyperlipidemic state is controversial and centers around whether there is an increased hepatic synthesis[1] of lipoproteins or a decrease in their catabolic rate.[2] A reasonable consensus is that the hyperlipidemia of nephrosis represents a complex pathophysiologic "adjustment" in which both an increased hepatic synthesis of lipoproteins, in combination with an absence of a comparable increase in lipoprotein catabolism, accounts for the resultant extreme hyperlipidemia.[3] A detailed discussion of the pathophysiologic mechanisms of abnormal lipid metabolism that occurs in the nephrotic syndrome and other renal diseases in humans can be found in Chapters 3 and 4 and in other recent reviews.[4]

DYSLIPIDEMIAS IN NON-NEPHROTIC PROGRESSIVE RENAL DISEASE

The major dyslipidemia of patients with chronic renal failure is an increase in plasma triglycerides, most probably secondary to reduced lipolysis of triglyceride-rich lipoproteins.[4] Triglycerides are markedly elevated (i.e., a range of 200 to 600 mg/dl) and are associated with a predominant increase in very low-density lipoprotein (VLDL). Intermediate-density lipoproteins can be elevated and the low-density lipoprotein (LDL) class exhibits an augmented triglyceride fraction.[5] In addition, circulating inhibitors of lipolytic enzymes may compound the hypertriglyceridemic abnormality.[6] In some patients with chronic renal failure, hypertriglyceridemia may be aggravated by obesity, diabetes mellitus, corticosteroids, and other pharmacologic agents. Increased triglyceride levels occur both before and after initiation of hemodialysis, and it has not been proven that hemodialysis will consistently mitigate the severity of the hypertriglyceridemia,[4] although with continuous ambulatory peritoneal dialysis the increased glucose and, thus, calorie diffusion across the peritoneal membrane can aggravate existing hypertriglyceridemia.

In the post-renal allograft setting, some patients manifest primarily hypertriglyceridemia, while others exhibit a predominant elevation in circulating cholesterol levels.[4] Corticosteroid treatment enhances the peripheral resistance to insulin action, thereby raising circulating insulin levels with a consequent increased VLDL synthesis. This purported mechanism can explain the elevation in circulating triglyceride levels; however, the elevation in cholesterol observed in some patients remains unexplained.

ANALOGY OF GLOMERULOSCLEROSIS TO ATHEROSCLEROSIS

Our laboratory, among others, has attempted to investigate the pathobiologic similarities between the atherosclerotic process and glomerulosclerosis, the latter recently becoming associated with the histochemical identification of macrophages, cholesterol, and cholesteryl esters in sclerotic glomeruli.[7] The atherosclerotic plaque bears many similarities to chronic inflammatory conditions, perhaps the most striking being the accumulation of macrophages in both situations. In atherogenesis, fatty streaks may develop into fibrous plaques with necrotic, lipid-filled macrophage cores surrounded by proliferating vascular smooth muscle cells (VSMCs) and increased amounts of connective tissue. However, in early lesions of fat-fed swine, Gerrity[8,9] showed that monocytes enter preatherosclerotic areas by penetrating the endothelium and take up cholesterol in the intima to become foam cells.

The importance of the monocyte/macrophage in the pathobiology of the atherosclerotic process has recently been reviewed by Steinberg and colleagues.[10,11] For many years, the foam cell was thought to arise from smooth muscle cells that had migrated into the intima and began to ingest lipids. More recent evidence has established that many foam cells are derived from circulating monocytes that adhere to the endothelium, penetrate into the subendothelial space, and there take up lipoproteins to become loaded with cholesteryl esters. The relative ineffectiveness of LDL in generating foam cells led to the speculation that some modified form of LDL might account for the atherogenicity of LDL. By combining the newer findings about the properties of oxidized LDL and the information available about the experimental pathology and cell biology of the macrophage, a hypothesis about the development of the fatty streak lesion based solely on the presence of elevated plasma LDL levels plus the oxidative modification of LDL within the arterial wall was constructed.[10,11] Basically, in the presence of a high plasma LDL concentration, the level of this lipoprotein is also increased in the intima of the vessel wall. Within the vessel wall, this native LDL is oxidized by either endothelial cells, macrophages, or smooth muscle cells contributing to the subsequent recruitment of circulating monocytes. Once within the arterial wall, the monocyte undergoes phenotypic modification and becomes a tissue macrophage. Its return to the plasma compartment is now inhibited by the oxidized LDL. This tissue macrophage now takes up oxidized LDL via the "scavenger" pathway and becomes a foam cell.[10,11]

As shown in Figure 6-1, the histologic similarities between a cholesterol-fed rat with a balloon-injured carotid artery (Fig. 6-1A) and a cholesterol-fed rat with a glomerulus manifesting the morphologic features of glomerulo-

A 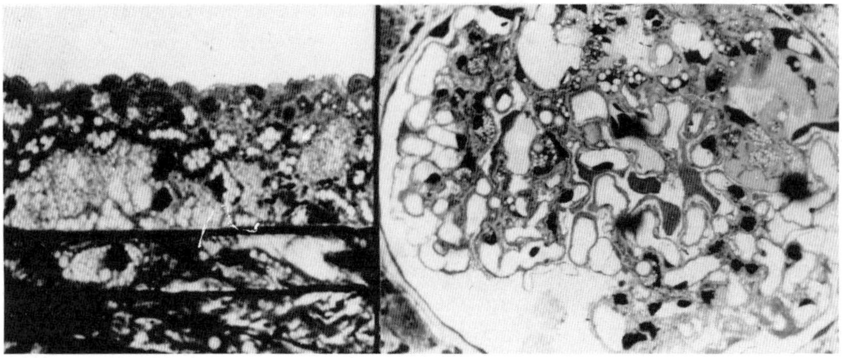 B

Fig. 6-1. (**A**) Light photomicrograph of a rat carotid artery 4 weeks after balloon injury and maintenance of the animal on a 4 percent cholesterol/1 percent cholic acid dietary supplement. Note the numerous intimal and medial foam cells, which have been subsequently histochemically identified as macrophages. (toluidine blue, ×600). (**B**) Light photomicrograph of a representative glomerulus from a rat that received puromycin aminonucleoside 18 weeks previously and was maintained on the same 4 percent cholesterol/1 percent cholic acid dietary supplement. Note the numerous foam cells within the glomerular mesangium, which, likewise, have been histochemically and immunohistochemically identified as macrophages. (Toluidine blue, ×650.) This composite photomicrograph depicts some of the analogous morphologic abnormalities that are common to both the atherosclerotic and glomerulosclerotic processes. A central feature to both processes is the influx of circulating monocytes into the vessel wall or mesangium to become lipid-laden foam cells.

sclerosis (Fig. 6-1B) are apparent. Both the intima and media of the artery contain numerous cholesterol-filled foam cells, whereas the corresponding area within the glomerulus, the mesangium, contains both increased amounts of foam cells and neutral lipid. In this analogy with the injured vessel wall, circulating peripheral blood monocytes can also infiltrate into damaged glomeruli, manifested clinically by albuminuria, and ingest cholesteryl ester to become foam cells; mesangial cells (MCs) proliferate in a fashion comparable to that of VSMCs in the arterial walls, and collagenous and noncollagenous extracellular matrix expansion takes place within the inflamed glomerular mesangium as in atheromatous arteries. Finally, both glomerulosclerosis and atherosclerosis are accelerated by the presence of hypercholesterolemia.[12–15]

Central to this analogy between the atherosclerosis and glomerulosclerosis processes is the origin and functional properties of the glomerular MCs (Table 6-1). The mesangium has been shown to contain at least two functionally distinct subgroups of cell types.[16] One cell type is smooth muscle-like and is probably derived from the paraglomerular mesoderm. Contractile glomerular MCs and VSMCs are closely related in terms of origin, microscopic anatomy and histochemistry, and contractility.[15] In addition, it has been demonstrated, with regard to the regulation of VSMC proliferation, that heparin inhibits the growth of this cell type in vivo and in vitro.[15]

Table 6-1. Similarities Between Vascular Smooth Muscle Cells and Mesangial Cells

Property	Vascular Smooth Muscle Cell	Contractile Mesangial Cell
Embryonic origin	Mesoderm	Paraglomerular mesoderm
Cytoskeletal components	Myosin, actin	Myosin, vimentin, desmin
Surface markers	AII receptors	AII receptors
Contractile response	AII, AVP, norepinephrine, PAF; calcium-dependent	AII, AVP, norepinephrine, PAF; calcium-dependent
Mitogens	IL-1, PDGF, and other growth factors	IL-1, PDGF, and other growth factors
Growth inhibitors	Anticoagulant and non-anticoagulant heparins	Anticoagulant and non-anticoagulant heparins
Products generated	Neutral proteinases	Neutral proteinases, reactive oxygen moieties

Abbreviations: AII, angiotensin II; AVP, arginine vasopressin; PAF, platelet-activating factor; IL-1, interleukin-1; PDGF, platelet-derived growth factor.

Castellot et al.[17] have also shown that heparin likewise inhibits glomerular MCs in culture, while in vivo both anticoagulant and non-anticoagulant heparin moieties are efficacious in decreasing the percentage of glomeruli with segmental MC proliferative areas in chronic aminonucleoside nephrosis.[18] Finally, contractile MCs share common properties with VSMCs, including angiotensin II receptors, calcium-dependent contractile response to several mediators (e.g., angiotensin II, arginine vasopressin, platelet-derived growth factor), and a proliferative response to platelet- and macrophage-derived products.[16] The second cell type is derived from the bone marrow. It constitutes approximately 3 to 5 percent of the total cells in the mesangium and is a macrophage. The macrophage-like MCs have been shown to bear Ia and common leukocyte antigens, which are both indicative of a hematopoietic derivation.[16]

HYPERCHOLESTEROLEMIA AS AN EXACERBATING FACTOR IN GLOMERULOSCLEROSIS

Clinical observations in patients suffering from the rare genetic disorder of lecithin-cholesterol-acyltransferase deficiency have provided preliminary patient evidence that perturbations in lipid metabolism can lead to abnormal lipoproteins and proteinuria.[19] Table 6-2 demonstrates the numerous animal studies that provide experimental support for the above-mentioned clinical findings and that serve as the basis to elucidate the pathobiologic mechanisms by which hypercholesterolemia can aggravate glomerular injury and, in certain models cause it. In normal, adult male Sprague-Dawley rats fed diets supplemented with either 3 percent cholesterol or 3 percent cholic acid and taurine for 2 to 80 weeks, significant hypercholesterolemia developed as early as 4 weeks after the start of the diet.[20] In addition, significant morphologic changes were observed in the aorta and kidney of rats 1 year of age and older in association with elevated plasma cholesterol levels in the dietary-supplemented groups.[20] Raised intimal plaques in the aorta were microscopically composed of large mononuclear cells with varying degrees of necrosis and calcification.[20] In the kidneys of the cholesterol- or cholic acid-fed rats, morphologic abnormalities were localized primarily in the glomeruli.[20] Sudanophilic droplets, hyalinosis, and segmental and/or global glomerulosclerosis lesions were present with moderate interstitial fibrosis.[20] These data demonstrate that in normal laboratory rats, glomerulosclerosis develops after a sufficiently long duration of cholesterol dietary supplementation.[20] In the normal guinea pig, Al-Shebeb and colleagues[21] demonstrated that a 2 percent dietary cholesterol supplement given over 70 days produced elevations in total cholesterol, free cholesterol, and phospholipid at 10, 30, and 70 days. The histologic appearance of animals fed the normal control diet revealed normal glomeruli, tubules, and vessels.[21] Thirty days after the cholesterol supplement was commenced, there was a variable increase in the mesangial matrix with mesangial hypercellularity that became even more pronounced by day 70.[21] Also, at day 70, glomeruli from the cholesterol-fed guinea pigs manifested a substantial increase in the number of nonspecific esterase-positive cells within the mesangium.[21] Although a causal role was not demonstrated, Al-Shebeb et al. hypothesized that an intraglomerular monocytic infiltration possibly participates in the development of a progressive glomerular lesion.[21] In an earlier study, French et al.[22] showed that guinea pigs made anemic and then fed a 1 percent cholesterol diet also developed a progressive glomerular lesion after 19 weeks of this dietary supplement, which was characterized by a thickening of the glomerular stalk in association with hyalin deposits in the vicinity of the juxtaglomerular apparatus.[22] In addition, macrophage-like cells were also occasionally seen within the hyalin deposits.[22] Our laboratory noted that feeding a 4 percent cholesterol/1 percent cholic acid dietary supplement to rats after initiating aminonucleoside nephrosis resulted in both a morphologic and functional exacerbation of this progressive glomerular disorder.[13] At the termination of the 18-week study, there was a significant elevation in proteinuria, a decline in whole-kidney glomerular filtration rate, and an elevation in serum urea nitrogen.[13] Nephrotic animals fed the cholesterol supplement had much more profound morphologic injury, as evidenced by a greater prevalence of mesangial foam cells within glomeruli (Fig. 6-2A) and a higher percentage of glomerulosclerosis lesions (Fig. 6-2B).[13] In addition, there was segmental

Table 6-2. Aggravating Effects of Dietary Cholesterol on Glomerular Injury

Model	Dietary Supplement Used	Outcome	References
Normal rat	3% cholesterol or 3% cholic acid with taurine	Aortic plaques and GS at 1 year after diet	20
Normal guinea pig	2% cholesterol	Proteinuria, mesangial expansion, increased glomerular monocytes by day 70 of diet	21
Anemic guinea pig	1% cholesterol	Enhanced GS and hyalin deposits in JGA region	22
Chronic PA nephrosis (rat)	4% cholesterol/1% cholic acid	Aggravation of GS, proteinuria, and a decrease in GFR with increased mesangial foam cells and neutral lipid	13
Normal, uninephrectomized, and two-kidney/one clip hypertensive rats	5% cholesterol	Augmented degree of GS in all three models after months on diet; clipped kidney in cholesterol-fed group only had an increased number of glomeruli with mesangial expansion	23
Normal and uninephrectomized rats	4% cholesterol/2% sodium cholate	Increased GS, cortical cholesteryl ester content, increased glomerular macrophages, and increased P_{GC} in cholesterol-fed nephrectomized rats	24
Autoimmune NZB/NZW mice with lupus nephritis	52% high-saturated fat diet (versus 4.5% fat)	Accelerated loss of renal function and premature mortality in high-fat group independent of changes in systemic immune parameters	25

Abbreviations: GS, glomerulosclerosis; JGA, juxtaglomerular apparatus; GFR, glomerular filtration rate; P_{GC}, glomerular capillary hydraulic pressure; PA, puromycin$_{GC}$ aminonucleoside.

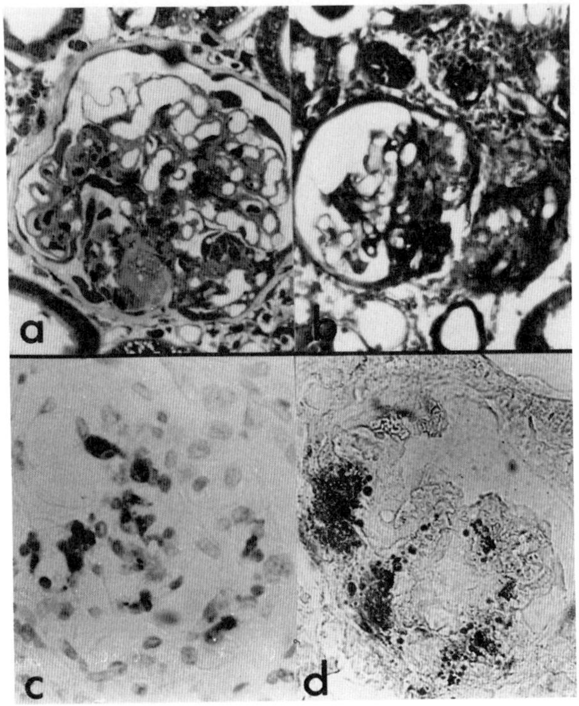

Fig. 6-2. Composite photomicrograph demonstrating representative glomeruli from rats 18 weeks after receiving a solitary intrajugular injection of puromycin aminonucleoside and being maintained for the entire course on a 4 percent cholesterol/1 percent cholic acid dietary supplement. (**A**) Epon-embedded, toluidine blue-stained glomerulus demonstrating a widely expanded glomerular mesangium filled with numerous vacuolated cells (foam cells). In addition, certain segments of the glomerulus show collapse of the capillary tuft with the presence of both amorphous material and foci of visceral epithelial cell blebbing. (**B**) Paraffin-embedded, periodic acid-Schiff-stained glomerulus with nearly global collapse of the capillary tuft, mesangial sclerosis, and a surrounding mononuclear cell tubulointerstitial infiltrate within areas of tubular atrophy and dilatation. (**C**) Paraffin-embedded immunoperoxidase staining with ED 1, a monoclonal mouse anti-rat monocyte/macrophage antibody. There is an increased number of glomerular macrophages within the mesangium, and the presence of these infiltrating cells correlates with the identification of the foam cells shown in Fig. A. (**D**) Frozen section of a glomerulus stained with Oil red O that reveals the presence of neutral lipid segmentally deposited within the glomerular mesangium. All of these morphologic changes within glomeruli of rats made nephrotic with puromycin aminonucleoside 18 weeks previously were significantly greater in animals that had been maintained on the dietary cholesterol supplement in contrast to normal rodent chow.[13]

deposition of neutral lipid within the mesangium, as detected by Oil red O staining on frozen tissue (Fig. 6-2D).[13] Using a mouse monoclonal antibody, ED 1, which is directed against rat monocyte/macrophage cytoplasmic antigens, the majority of mesangial foam cells have been shown to be macrophages (Fig. 6-2C).

What pathobiologic mechanism(s) mediates the deleterious effects of cholesterol feeding? Grone et al.[23] have recently reported that in rats fed a high-fat/high-cholesterol diet, the incidence of glomerulosclerosis increased significantly in normal animals after 6 months. Glomerulosclerosis induced by the high-fat diet was aggravated by unilateral nephrectomy and systemic arterial hypertension, suggesting that altered glomerular hemodynamics may contribute to the sclerosis process.[23] Kasiske et al.,[24] using stop-flow micropuncture techniques, have shown that high cholesterol intake had no statistically significant effect on either single-nephron glomerular filtration rate or single-nephron plasma flow, but had dramatic effects on intraglomerular pressures. Both the glomerular capillary hydraulic pressure and the transcapillary hydraulic pressure difference were elevated in cholesterol-fed rats.[24] There was no statistically significant effect of high cholesterol intake on either the afferent or efferent arteriolar resistances.[24] There was a small, but statistically significant, effect of high cholesterol intake on mean arterial blood pressure at the time of micropuncture; however, none of the cholesterol-fed rats were hypertensive.[24] Kelley and Izui[25] found that a lipid-rich diet caused an earlier death in NZBxW mice with autoimmune lupus nephritis by promoting renal disease. Ultrastructural and histochemical evaluation of glomeruli from nephritic mice fed the lipid-rich diet demonstrated lipid in the peripheral capillary loops and mesangium adjacent to immune deposits.[25] The accelerated renal disease was not related to changes in systemic immune parameters, suggesting that the high lipid consumption hastened renal deterioration by some noxious local "lipid-glomerular capillary" interaction and not by a systemic influence on the immune response.[25]

Other intrarenal adaptations to the hypercholesterolemic milieu have been recently examined and will be presented in other chapters in this volume. These potentially deleterious effects of cholesterol feeding include biochemical alterations in the renal cortical lipid content[24] and evidence that demonstrates both uptake and toxicity of LDLs on cultured rat glomerular cells.[26]

LIPID-LOWERING AGENT INTERVENTION STUDIES

A variety of lipid-lowering agents, including clofibric acid, lovastatin, and cholestyramine, have been shown to reduce circulating lipids and preserve renal structure in models of progressive renal disease.[27–31] A detailed discussion of these data can be found in Chapter 7. Collectively, these data show that in both hypertensive and normotensive models of renal disease, lipids play an important role in nephron injury. In addition, they also suggest that the lipid-lowering effects of these agents are critical to the prevention of renal disease, since cholestyramine exerts its effects without systemic absorption, in contrast to the other hypolipidemic agents that have been used.

THE GLOMERULAR MACROPHAGE AS A POTENTIAL EFFECTOR MECHANISM FOR THE DELETERIOUS EFFECT OF HYPERCHOLESTEROLEMIA ON GLOMERULAR INJURY

An increasing amount of experimental evidence suggests that changes in lipoproteins can alter the functional properties of the monocyte/macrophage, thereby enabling this cell to participate as a mediator in the atherosclerotic process.[10,11] Hartung et al.[32] have shown that modified LDL moieties augment murine resident peritoneal macrophage eicosanoid production and release of both the superoxide anion and hydrogen peroxide. It has also been recently shown[33] that cholesterol-rich macrophages synthesize and secrete significantly more 12-hydroxyeicosatetraenoic acid from exogenous arachidonic acid. Also, mouse peritoneal macrophages enriched in cholesterol by acetyl-LDL, a modified LDL moiety, release less prostacyclin and prostaglandin E_2 from endogenous stores of arachidonic acid compared with the amount of prostaglandins released by unmodified cells.[34] In association with these above-mentioned alterations in eicosanoid biosynthesis in relation to cholesterol ingestion by macrophages, there was a relative preservation in the amount of thromboxane B_2 produced.[34] These alterations in the generation in the cyclooxygenase and lipoxygenase derivatives of arachidonic acid may promote hemodynamic changes, alter the interactions between a variety of immune cells, recruit other inflammatory cells to affected tissue sites, and enhance vascular permeability. Monocyte/macrophage stimulation of the scavenger receptors expressed on the surface of these cells could initiate a series of reactions that trigger release of mediators that either directly or indirectly impose injury on target vasculature. Rogers et al.[35] have demonstrated that in normal rats, alimentary hypercholesterolemia augmented several macrophage functions, including adhesion to endothelium and secretion of chemotactic and mitogenic substances. These investigators postulated that dietary cholesterol may accelerate atherosclerosis by inducing specific changes in properties of circulating monocytes and intimal macrophages.[35] Kelley et al.,[36] using isolated human peripheral blood monocytes, have shown that LDL and acetylated LDL stimulated secretion of beta-glucouronidase activity into the culture medium. Incubation of monocytes with these LDL moieties in the presence of inhibitors of arachidonate metabolism resulted in a significant decrease in β-glucouronidase activity, indicating a role for products of arachidonate metabolism in the activation of monocytes by lipoproteins.[36] Thus, these representative studies demonstrate the complex cell biologic alterations that develop in monocytes/macrophages as a result of cholesterol feeding in vivo and the in vitro addition of native and modified LDL to macrophages in cell culture.

As alluded to previously, one can draw many analogies between the processes of atherosclerosis and glomerulosclerosis. Recently, data have emerged to suggest that the lipid abnormalities associated with nephrosis can also result in interactions with monocytes/macrophages. Our laboratory[37] has shown that superimposing alimentary hypercholesterolemia onto the nephrotic state (i.e., 2 weeks after the administration of puromycin aminonucleoside [PA]) significantly increased the glomerular macrophage number as well as peritoneal macrophage phagocytosis and basal thromboxane B_2 production, in contrast to normal rats with only alimentary hypercholesterolemia and nephrotic animals with only the consequent increments in fasting total serum cholesterol and triglycerides resulting from the glomerular abnormality. These latter nephrotic animals, which were maintained on a standard laboratory chow, also had a significantly greater glomerular macrophage number when compared with a normal cohort of rats on the same diet.[37] These data suggest that there are individual causal roles for both dietary hypercholesterolemia and the nephrotic state to increase the glomerular macrophage number as well as a synergistic effect between these two perturbations in lipid metabolism. Subsequently, this observation whereby dietary hypercholesterolemia increases glomerular macrophage number in the rat has been confirmed by others.[24] Coupled with our earlier findings, which demonstrated an exacerbation of chronic aminonucleoside nephrosis by cholesterol feeding,[13] these more recent data showing a cholesterol-mediated augmentation in the glomerular macrophage number suggest that the monocyte/macrophage may be the effector cell through which hypercholesterolemia and alterations in lipoprotein metabolism mediate the propagation of initial glomerular injury to glomerulosclerosis. In analogy to the atherosclerotic process, the hypercholesterolemia of nephrosis, with alterations in circulating lipoproteins, could favor the influx of circulating monocytes into an injured glomerular mesangium and foster the activation of these glomerular macrophages as they become foam cells. This hypothetical process could mediate a number of proinflammatory events, such as release of proteinases, generation of oxygen-free radicals, elaboration of monokines and other growth factors, and production of other chemotactic substances that collectively impact on other resident glomerular cell types (notably, the contractile glomerular mesangial cell) in an adverse way and initiate a cascade of additional events that will culminate in mesangial cell proliferation and matrix overproduction, which are forerunners of glomerulosclerosis.

Recent experimental investigations involving nonimmune models of glomerular disease are beginning to confirm an important role for glomerular macrophages in participating in the glomerulosclerotic process. Matsumoto and Atkins[38] isolated glomeruli from PA nephrotic rats and grew them in tissue culture. The glomeruli from nephrotic rats had significantly greater numbers of macrophages present in the outgrowths when compared with saline control animals.[38] In addition, the number of glomerular macrophages in the outgrowths and the release of interleukin-1 into the glomerular supernatant were greatest from glomeruli isolated from rats at day 10 of the disease course,[38] which represents a time point when there is a surge in infiltrating glomerular macrophages in temporal association with peak albuminuria and the consequent hypercholesterolemia.[37–40] Likewise, our laboratory has demonstrated in vivo a statistically significant increase in the immunohistochemical expression of interleukin-1-α and tumor necrosis factor-α in cells within the glomerular mesangium during acute aminonucleoside nephrosis (i.e., 2 weeks after PA).[41] Double immunohistochemical labeling revealed that these intracellular cytokines were present within macrophages.[41] Saito and Atkins[42] have also recently shown, using quantitative immunohistochemical analysis of cell types within the glomeruli and tubulointerstitium during the course of chronic aminonucleoside nephrosis, that glomerulosclerotic lesions in this model are related to the presence of macrophages during the early phase of the disease course.

The question remains whether any of this experimental data have any bearing to focal glomerulosclerosis in humans. Magil and Cohen[43] evaluated biopsy specimens from 13 patients with focal glomerulosclerosis by combined histochemical and ultrastructural means. Over 70 percent of glomeruli with segmental sclerosis lesions contained monocytes and/or foam cells, whereas under 10 percent of glomeruli without the lesions demonstrated these cells.[43] In glomeruli with sclerotic lesions, foam cells predominated over monocytes.[43] Neutral lipid was observed segmentally and focally in the vast majority of biopsy specimens with focal glomerulosclerosis.[43] In very early lesions, characterized ultrastructurally by predominantly visceral epithelial cell changes without glomerular tuft alterations, the infiltrating intraglomerular cells were mainly monocytic with few foam cells.[43] With progressive lesions, the relative number of monocytes declined, but foam cells were observed more frequently.[43] Although this clinical study did not demonstrate a causal effect between glomerular macrophages and progressive glomerulosclerosis, these data[43] do suggest that monocytes and their derivatives are involved in the development of focal glomerulosclerosis.

To confirm whether glomerular macrophages may be an effector mechanism that mediates the noxious effects of hypercholesterolemia on progressive glomerular disease, our laboratory has recently completed a number of macrophage depletion studies. With the observation that an essential fatty acid-deficient (EFAD) diet depletes normal rat glomeruli of resident macrophages,[44] we examined the effect of this diet in the progressive chronic aminonucleoside nephrosis model.[40] Rats were fed an EFAD diet for 8 weeks and then received a single intravenous injection of PA. Nephrotic animals were continued on this EFAD diet for *only* an additional 4 weeks, which corresponds temporally to the duration of acute nephrosis in this model, and then were transferred to standard laboratory chow for the duration of the 18-week study. The recurrent albuminuria and development of glomerulosclerosis that characterize the later stage of this model were significantly blunted in EFAD rats. These late beneficial outcomes (i.e., 18 weeks after PA) were associated with significant reductions in the glomerular macrophage number in EFAD rats during acute nephrosis (i.e., 2 weeks after PA), suggesting that modulation of this parameter during acute nephrosis could have important long-term effects on the course of progressive glomerular injury. It should be emphasized that during the acute albuminuric phase in this study, both the EFAD and control diet nephrotic groups had equivalent degrees of fasting hypercholesterolemia and hypertriglyceridemia. Although the hypercholesterolemia accompanying glomerular injury is an exacerbating factor favoring the progression to glomerulosclerosis, this recent data[40] demonstrates that if the glomerular macrophage number is lowered during acute PA nephrosis independent of any significant change in the fasting total cholesterol levels, then protection can be conferred against progressive glomerulopathy. These results suggest that the presence of infiltrating macrophages within the mesangium soon after nonimmune glomerular injury plays an important role in the propagation of initial glomerular injury to glomerulosclerosis. Thus, the augmented number of glomerular macrophages in the settings of nephrotic-associated or dietary-induced hypercholesterolemia underscores this cell type as a potential effector mechanism in progressive glomerular disease. Further support for this determining role of glomerular macrophages in progressive chronic aminonucleoside nephrosis has recently been obtained[45] using another interventional maneuver that reduces the glomerular macrophage number, whole body x-irradiation.[46] A single sublethal dose of whole body x-irradiation with bilateral kidney shielding was administered within days after PA delivery to produce its maximal effect in temporal concordance with the acute albuminuric phase of PA nephrosis and the concurrent surge in infiltrating glomerular macrophages.[45] Without altering the levels of albuminuria or circulating fasting lipid levels during acute nephrosis, the irradiated PA nephrotic rats had significantly fewer glomerular macrophages 11 days after PA in contrast to sham-irradiated nephrotic control rats.[45] In addition, at 8, 12, and 18 weeks after PA, the PA nephrotic rats that received the solitary dose of irradiation had no recurrent albuminuria, in contrast to the PA nephrotic rats that were sham-irradiated and manifested the typical serial significant increments in albuminuria.[45] Normal rats, which were irradiated, manifested no renal dysfunction or albuminuria as a consequence of the radiation. At the conclusion of this 18-week study, both irradiated normal and PA nephrotic rats had 1 to 2 percent of glomeruli exhibiting glomerulosclerosis lesions in comparison to a significantly greater percentage (13 percent) of glomeruli exhibiting either segmental or globally sclerotic lesions in the sham-irradiated PA nephrotic group.[45] These data confirm the beneficial results obtained with an EFAD diet[40] in ameliorating the progression of chronic aminonucleoside nephrosis by reducing the glomerular macrophage number during acute nephrosis.

SUMMARY

As shown in Figure 6-3, the hypercholesterolemia accompanying nephrosis, cholesterol feeding, or both could mediate, in part, the transition from initial glomerular injury to glomerulosclerosis by an increase in the glomerular macrophage number. In addition, hypercholesterolemia could be associated with increased circulating LDLs and/or other lipoprotein abnormalities as well as a generalized aggravation of atherosclerosis systemically with this disorder's attendant complications. The elevation in the glomerular macrophage number could potentially result in an augmented release of macrophage-derived oxygen-free radicals and neutral proteinases, as well as perturb glomerular eicosanoid metabolism, the latter of which may be a modulating factor in the development of glomerulosclerosis.[47] All of these phenomena are proinflammatory and could adversely impact on the contractile glomerular mesangial cell and/or matrix, resulting in cell injury and matrix overproduction, both of which are harbingers of glomerulosclerosis. In addition, infiltrating macrophages gaining access to the glomerular mesangium in temporal association with the glomerular injury and augmented by the accompanying secondary hypercholesterolemia could be elaborating peptide growth factors or monokines (e.g., interleukin-1, tumor necrosis factor, platelet-derived growth factor, etc.), which may further aggravate this glomerular mesangial proinflammatory state by directly initiating perturbations in contractile mesangial cell function. These include increased proliferation,[48,49] augmented neutral proteinase (e.g., gelatinase) production,[50] and release of mesangial cell-derived toxic oxygen species.[51] Interventional strategies during acute nephrosis that decrease the glomerular macrophage number, such as EFAD[40] and whole body x-irradiation,[45] have been demonstrated to confer protection against progressive glomerulopathy in the experimental nephrotic syndrome produced by PA. These recent data further underscore the important role of the glomerular macrophage in propagating initial glomerular injury to glomerulosclerosis.

In conclusion, this review has attempted to demonstrate the importance of hypercholesterolemia in the pathogenesis of progressive glomerular injury. In addition, the emerging concept that an increased traffic of the monocyte/macrophage through the glomerular mesangium may be one potential effector mechanism by which hypercholesterolemia exacerbates glomerular injury has been presented. Clearly, additional studies are needed to confirm the paradigm that suggests analogous pathobiologic processes in both glomerulosclerosis and atherosclerosis.

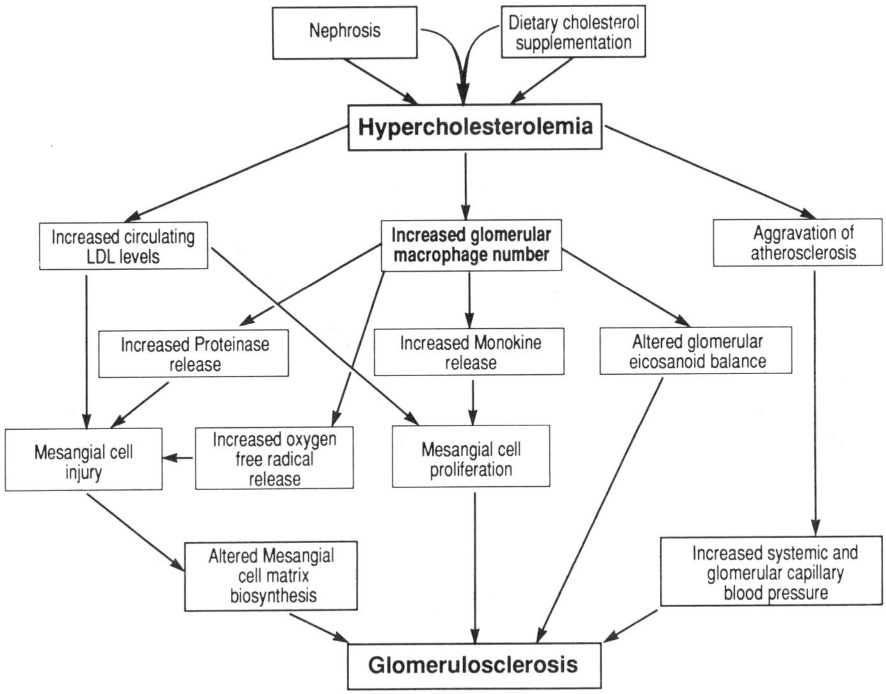

Fig. 6-3. Schematic diagram depicting how the hypercholesterolemia associated with nephrosis, dietary cholesterol supplementation, or both could mediate, in part, the progression of initial glomerular injury to glomerulosclerosis. Central to this paradigm is an increase in the glomerular macrophage number produced by the hypercholesterolemia. The increase in the number of glomerular macrophages could theoretically produce an increased release of macrophage-derived products, including oxygen-free radicals, neutral proteinases, and certain eicosanoids. These products are proinflammatory and could produce mesangial cell injury and an altered mesangial cell matrix biosynthesis. Infiltrating macrophages could elaborate peptide growth factors or monokines, which can cause mesangial cell proliferation and, although not shown by arrows, can stimulate mesangial cells to produce neutral proteinases and release reactive oxygen moieties, causing further glomerular injury. Finally, hypercholesterolemia could alter circulating lipoprotein levels, which could adversely effect contractile glomerular mesangial cells on gaining access to the mesangium. Hypercholesterolemia could exacerbate the atherosclerotic process systemically, which could raise arterial pressure in both large vessels and the glomerular microcirculation. Although no arrow is shown, there is some recent experimental evidence linking dietary hypercholesterolemia to an elevated glomerular capillary hydraulic pressure in the absence of systemic hypertension (see text for details).

ACKNOWLEDGMENTS

Experimental studies referred to in this chapter were funded by National Institutes of Health Grants AM 13132, DK 38394, and DK 40839, as well as a Clinician-Scientist Award and Research Grants-in-Aid (Massachusetts and Pennsylvania affiliates) from the American Heart Association.

REFERENCES

1. Marsh JB, Drabkin DL: Experimental reconstruction of metabolic pattern of lipid nephrosis: key role of hepatic protein synthesis in hyperlipemia. Metabolism 9:946, 1960
2. Staprans I, Garon SJ, Hopper J, Felts JM: Characterization of glycosaminoglycans in urine from patients with nephrotic syndrome and control subjects, and their effects on lipoprotein lipase. Biochim Biophys Acta 678:414, 1981
3. Marsh JB: Lipoprotein metabolism in experimental nephrosis. J Lipid Res 25:1619, 1984
4. Grundy SM: Management of hyperlipidemia of kidney disease. Kidney Int 37:847, 1990
5. Nestel PJ, Fidge NH, Tan MH: Increased lipoprotein-remnant formation in chronic renal failure. N Engl J Med 307:239, 1982
6. Murase T, Cattran DL, Rubenstein B, Steinder G: Inhibition of lipoprotein lipase by uremic plasma. A possible cause for hypertriglyceridemia. Metabolism 24:1279, 1975
7. Grond J, van Goor H, Erkelens DW, Elema JD: Glomerular sclerotic lesions in the rat: histochemical analysis of their macromolecular and cellular composition. Virchows Arch [B] 51:521, 1986
8. Gerrity RG: The role of the monocyte in atherogenesis. I. Transition from blood-borne monocytes into foam cells in fatty lesions. Am J Pathol 103:181, 1981
9. Gerrity RC: The role of the monocyte in atherogenesis. II. Migration of foam cells from atherosclerotic lesions. Am J Pathol 103:191, 1981
10. Steinberg D: Lipoproteins and the pathogenesis of atherosclerosis. Circulation 76:508, 1987
11. Steinberg D, Parthasarathy S, Carew TE et al: Beyond cholesterol: modifications of low-density lipoprotein that increase its atherogenicity. N Engl J Med 320:915, 1989
12. Moorhead JR, El-Nahas M, Chan MK, Varghese Z: Lipid nephrotoxicity in chronic glomerular and tubulo-interstitial disease. Lancet 2:1309, 1982
13. Diamond JR, Karnovsky MJ: Exacerbation of chronic aminonucleoside nephrosis by dietary cholesterol supplementation. Kidney Int 32:671, 1987
14. Keane WF, Kasiske BL, O'Donnell MP: Lipids and progressive glomerulosclerosis: a model analogous to atherosclerosis. Am J Nephrol 8:261, 1988
15. Diamond JR, Karnovsky MJ: Focal and segmental glomerulosclerosis: analogies to atherosclerosis. Kidney Int 33:917, 1988
16. Striker GE, Striker LJ: Glomerular cell culture. Lab Invest 53:122, 1985
17. Castellot JJ, Hoover RL, Harper PA, Karnovsky MJ: Heparin and glomerular epithelial cell-secreted heparin-like species inhibit mesangial cell proliferation. Am J Pathol 120:427, 1985

18. Diamond JR, Karnovsky MJ: Nonanticoagulant protective effects of heparin in chronic aminonucleoside nephrosis. Renal Physiol 9:366, 1986
19. Norum KR: Familial lecithin cholesterol acyltransferase deficiency. p. 297. In Miller NE, Miller GJ (eds): Clinical and Metabolic Aspects of High-Density Lipoproteins. Amsterdam, Elsevier Science Publishing, 1984
20. Peric-Golia L, Peric-Golia M: Aortic and renal lesions in hypercholesterolemic adult, male, virgin Sprague-Dawley rats. Atherosclerosis 46:57, 1983
21. Al-Shebeb T, Frohlich J, Magil AB: Glomerular disease in hypercholesterolemia guinea pigs: a pathogenetic study. Kidney Int 33:498, 1988
22. French SW, Yamanaka W, Ostwald R: Dietary induced glomerulosclerosis in the guinea pig. Arch Pathol 83:204, 1967
23. Grone H, Walli A, Grone E et al: Induction of glomerulosclerosis by dietary lipids. Lab Invest 60:433, 1989
24. Kasiske BL, O'Donnell MP, Schmitz PG et al: Renal injury of diet-induced hypercholesterolemia in rats. Kidney Int 37:880, 1990
25. Kelley VE, Izui S: Enriched lipid diet accelerates lupus nephritis in NZB x W mice. Am J Pathol 111:288, 1983
26. Wasserman JA, Santiago A, Rifici H et al: Interactions of low density lipoprotein with rat mesangial cells. Kidney Int 34:1168, 1989
27. Kasiske BL, O'Donnell MP, Garvis WJ, Keane WF: Pharmacologic treatment of hyperlipidemia reduces glomerular injury in rat 5/6 nephrectomy model of chronic renal failure. Circ Res 62:367, 1988
28. Golper TA, Feingold KR, Fulford MH, Siperstein MD: The role of circulating mevalonate in nephrotic hypercholesterolemia in the rat. J Lipid Res 27:1044, 1986
29. Kasiske BL, O'Donnell MP, Cleary MP, Keane WF: Treatment of hyperlipidemia reduces glomerular injury in obese Zucker rats. Kidney Int 33:667, 1988
30. Harris KPG, Purkerson ML, Yates J, Klahr S: Lovastatin ameliorates the development of glomerulosclerosis and uremia in experimental nephrotic syndrome. Am J Kidney Dis 15:16, 1990
31. Diamond JR, Hanchak NA, McCarter MD, Karnovsky MJ: Cholestyramine resin ameliorates chronic aminonucleoside nephrosis. Am J Clin Nutr 51:606, 1990
32. Hartung HP, Kladetzky RG, Melnik B, Hennerici M: Stimulation of the scavenger receptor on monocytes-macrophages evokes release of arachidonic acid mediators and reduced oxygen species. Lab Invest 55:209, 1986
33. Mathur SN, Field FJ, Spector AA, Armstrong ML: Increased production of lipoxygenase products by cholesterol-rich mouse macrophages. Biochim Biophys Acta 837:13, 1985
34. Mathur SN, Albright E, Field FJ: Decreased prostaglandin production by cholesterol-rich macrophages. J Lipid Res 30:1385, 1989
35. Rogers KA, Hoover RL, Castellot JJ et al: Dietary cholesterol-induced changes in macrophage characteristics: relationship to atherosclerosis. Am J Pathol 125:284, 1986
36. Kelley JL, Rozek MM, Suenram CE, Schwartz CJ: Activation of human peripheral blood monocytes by lipoproteins. Am J Pathol 130:223, 1988
37. Diamond JR, Pesek I, McCarter MD, Karnovsky MJ: Altered functional characteristics of rat macrophages during nephrosis: synergistic effects of hypercholesterolemia. Am J Pathol 135:711, 1989
38. Matsumoto K, Atkins RC: Glomerular cells and macrophages in the progression

of experimental focal and segmental glomerulosclerosis. Am J Pathol 134:933, 1989

39. Schreiner GF, Cotran RS, Unanue ER: Modulation of Ia and leukocyte common antigen expression in rat glomeruli during the course of glomerulonephritis and aminonucleoside nephrosis. Lab Invest 51:524, 1984
40. Diamond JR, Pesek I, Ruggieri S, Karnovsky MJ: Essential fatty acid deficiency during acute puromycin nephrosis ameliorates late renal injury. Am J Physiol 257:798, 1989
41. Diamond JR, Pesek I: Immunohistochemical demonstration of tumor necrosis factor and interleukin-1 during acute aminonucleoside nephrosis. Modulation by an essential fatty acid deficient. Lab Invest 64:21, 1991
42. Saito T, Atkins RC: Contribution of mononuclear leucocytes to the progression of experimental focal glomerular sclerosis. Kidney Int 37:1076, 1990
43. Magil AB, Cohen AH: Monocytes and focal glomerulosclerosis. Lab Invest 61:404, 1989
44. Lefkowith JB, Schreiner G: Essential fatty acid deficiency depletes rat glomeruli of resident macrophages and inhibits angiotensin II-induced eicosanoid synthesis. J Clin Invest 80:947, 1987
45. Diamond JR, Pesek-Diamond I: Sublethal X-irradiation during acute puromycin nephrosis prevents late renal injury: role of glomerular macrophages. Am J Physiol 260:F779, 1991
46. Schreiner GF, Unanue ER: Origin of the rat mesangial phagocyte and its expression of the leukocyte common antigen. Lab Invest 51:515, 1984
47. Salvati P, Ferti C, Ferrario RG et al: Role of enhanced glomerular synthesis of thromboxane A_2 in progressive kidney disease. Kidney Int 38:447, 1990
48. Lovett DH, Ryan JL, Sterzel RB: Stimulation of rat mesangial cell proliferation by macrophage interleukin-1. J Immunol 131:2830, 1983
49. Lovett DH, Szamel M, Ryan JL et al: Interleukin 1 and the glomerular mesangium. I. Purification and characterization of a mesangial cell-derived autogrowth factor. J Immunol 136:3700, 1986
50. Martin J, Lovett DH, Gemsa D et al: Enhancement of glomerular mesangial cell neutral proteinase secretion by macrophages: role of interleukin 1. J Immunol 137:525, 1986
51. Radeke HH, Meier B, Topley N et al: Interleukin-1 and tumor necrosis factor induce oxygen radical production in mesangial cell. Kidney Int 37:767, 1990

7

Dietary and Pharmacologic Manipulations of Lipids

Impact on Progression of Experimental Renal Disease

Michael P. O'Donnell
Paul G. Schmitz

INTRODUCTION

There is increasing evidence that lipids are independently important in the pathogenesis of chronic, progressive renal disease. The bulk of this evidence stems from studies in experimental models of renal disease, in which either pharmacologic or dietary maneuvers have been used to manipulate circulating lipids and lipoproteins. The purpose of this review is to summarize the effects of such maneuvers on renal structure and function in experimental and clinical renal disease. The review will focus primarily on the actions of various lipid-lowering drugs and dietary lipid modifications on glomerular filtration rate (GFR), glomerular hemodynamics, proteinuria, and glomerular histology in experimental models of progressive renal disease. Also to be discussed in brief will be the effects of hypolipidemic maneuvers on renal function and proteinuria in clinical renal disease.

ASSOCIATIONS BETWEEN HYPERLIPIDEMIA AND RENAL DISEASE

Renal lipid deposits have been recognized in a variety of renal diseases. Investigators have noted renal fatty deposits in patients with Bright's disease, nephrotic syndrome, diabetic nephropathy, and chronic renal failure.[1–8] Recently, much attention has been directed toward the potential nephrotoxicity of lipids.[8–10] Indeed, it has been suggested that lipid participation in the development of glomerulosclerosis may parallel the well-described lipid involvement in the pathogenesis of atherosclerosis.[8,10]

Clinically, alterations in circulating lipids and lipoproteins are frequently associated with renal disease. In particular, type 4 hypertriglyceridemia and type 2b combined hyperlipidemia have been frequently found in patients with nephrotic syndrome, analgesic nephropathy, and chronic renal failure, as well as in patients with renal allografts.[11] The role of lipids in the pathogenesis and progression of clinical renal disease is unknown. Studies in experimental models of progressive renal disease, however, strongly suggest an important role of lipids in the initiation and progression of glomerular injury.[12]

Analogous to the clinical situation, hyperlipidemia is frequently associated with experimental renal disease. In particular, several models of non-immune-mediated progressive glomerular injury are characterized by elevated serum cholesterol and triglycerides (Table 7-1). In the rat remnant kidney model of chronic renal failure, for example, hypercholesterolemia and hypertriglyceridemia occur in association with renal insufficiency, marked albuminuria, and accelerated development of glomerulosclerosis.[13]

Hyperlipidemia and glomerular disease are also found in several animal models of diabetes mellitus. Streptozotocin- or alloxan-treated rats have been frequently studied as a model of human type I diabetes, and these rats, particularly with poor metabolic control, demonstrate hyperlipidemia, al-

Table 7-1. Experimental Models of Non-Immune-Mediated Renal Disease Characterized by Hyperlipidemia

Renal ablation (remnant kidney; rats)
Diabetes mellitus
Streptozotocin-treated rats
Obese Zucker rats
Obese spontaneously hypertensive rats (obese SHR)
SHR/N-corpulent rats
db/db mice
Chinese hamsters
Nephrotic syndrome
Puromycin aminonucleoside (rats)
Adriamycin (rats)
Hypertension (Dahl salt-sensitive rats)
Diet-induced hypercholesterolemia (rats, rabbits, guinea pigs)
Spontaneously hypercholesterolemic rats

buminuria, and glomerulosclerosis.[14–16] Hyperlipidemia and glomerular structural changes have also been described in several rodent models of spontaneous diabetes, including Chinese hamsters,[14,17] obese Zucker rats,[18,19] obese spontaneous hypertensive rats (obese SHR),[14,17,20] SHR/N-corpulent rats,[17,21,22] and db/db mice.[14,17,23,24]

Experimental nephrotic syndrome produced in rats by the administration of puromycin aminonucleoside (PAN) or Adriamycin is characterized by massive proteinuria, hypoalbuminemia, and hyperlipidemia.[25–28] Glomerular histologic changes that develop in these nephrotic rats qualitatively resemble structural alterations in human nephrotic syndrome.[25,26] Moreover, nephrotic rats develop renal insufficiency.[27,28]

Glomerular disease and hyperlipidemia have been described in experimental hypertension. Prehypertensive Dahl salt-sensitive (S) rats have elevated serum triglycerides and increased urine albumin excretion.[29,31] When Dahl S rats are fed a high-salt diet, hypertension and marked increases in serum triglycerides, serum cholesterol, and albuminuria develop.[29–31] In one study, focal glomerulosclerosis (FGS) was detected in Dahl S rats as early as 15 weeks of age, and, at 30 weeks of age, approximately one-third of glomeruli in hypertensive Dahl S rats exhibited FGS.[29]

Glomerular injury also develops in SHR, although more slowly than in Dahl S rats.[32] Although normal serum lipid levels have been reported in SHR,[29] these rats do exhibit glucose intolerance and peripheral insulin resistance, which might alter lipid metabolism.[33,34]

Elevated serum cholesterol and triglycerides, proteinuria, and glomerulosclerosis have also been found in spontaneously hypercholesterolemic rats.[35]

Thus, in many experimental models, hyperlipidemia occurs in association with renal disease. Importantly, hyperlipidemia is frequently present prior to the development of marked albuminuria and glomerulosclerosis. Pharmacologic and dietary maneuvers, described in the next section, have been shown to lower circulating lipids and reduce glomerular injury in several of these models, thus providing strong evidence that lipids participate in the pathogenesis of glomerular disease.

An association between hyperlipidemia and experimental renal disease, however, is not universal. Several exceptions have been described. In the BB rat, a model of spontaneous insulin-dependent diabetes, hyperlipidemia is present,[14] but only minimal glomerular injury, primarily basement membrane thickening, has been described.[36,37] Nagase analbuminemic rats develop marked hyperlipidemia secondary to an absence of circulating albumin. Glomerular disease is normally minimal in these rats, although it can be accentuated by maneuvers such as subtotal nephrectomy.[38] Lastly, Watanabe heritable hyperlipidemic (WHHL) rabbits, which lack peripheral low-density lipoprotein (LDL) receptors and are markedly hypercholesterolemic, do not develop significant renal disease.[39,40] Interestingly, rabbits fed a high-cholesterol diet, in which serum total cholesterol levels are similar to those in WHHL rabbits, do develop severe glomerulosclerosis.[39,41] These findings suggest that an increase in lipid levels alone may not be sufficient to induce glomerular injury, but that other factors associated with hyperlipidemia, such as alterations in lipoprotein composition, may be required.

Renal injury can be induced not only in rabbits but also in other species of laboratory animals by feeding diets high in cholesterol or saturated fat. Diet-induced hyperlipidemia has been shown in rats and guinea pigs to be associated with albuminuria and accelerated development of glomerulosclerosis.[41–46] While these models of diet-induced hyperlipidemia and renal disease are certainly useful to study potential mechanisms by which lipids can injure the glomerulus, it should be noted that primary hyperlipidemias in patients do not necessarily cause renal disease. Patients with familial hypercholesterolemia, for example, do not appear to have increased risk of elevated urine albumin excretion or renal insufficiency, despite very high levels of circulating cholesterol.[40] On the other hand, there are reports of glomerular injury in patients with type III hyperlipidemia or a form of hyperlipidemia similar to type III hyperlipoproteinemia.[47,48]

PHARMACOLOGIC TREATMENT OF HYPERLIPIDEMIA IN RENAL DISEASE: EFFECTS ON RENAL STRUCTURE AND FUNCTION

Fibrates

Studies designed to examine the effects of lipid-lowering agents on the course of experimental renal disease were first performed by Edwards and co-workers in the late 1970s.[11,49,50] In those studies, nephrotic (PAN) rats that were fed a sucrose-lard diet developed marked hyperlipidemia and exaggerated renal disease, characterized by proteinuria, a high incidence of FGS, and reductions in GFR and renal plasma flow (RPF). Moreover, PAN rats fed the sucrose–lard diet had increased mortality compared with PAN rats fed standard laboratory chow.

Addition of the fibrate derivative halofenate to the sucrose-lard diet largely prevented glomerular disease in nephrotic rats. Halofenate-treated PAN rats had near normal plasma lipid levels, as well as normal values of GFR, RPF, and urine protein excretion.[49,50] Morphologically, renal lesions in halofenate-treated PAN rats were substantially less than in untreated PAN rats, and mortality was not different from that in PAN rats fed standard chow.

Subsequent studies have demonstrated beneficial effects of another fibrate, clofibric acid, in other experimental models of chronic renal disease. Clofibric acid is the pharmacologically active form of the hypolipidemic agent clofibrate. In separate studies, Kasiske and co-workers found that clofibric acid, administered subcutaneously once a day, reduced albuminuria and renal structural injury in subtotally ($\frac{5}{6}$) nephrectomized rats and obese Zucker rats.[13,19] Compared with untreated $\frac{5}{6}$ nephrectomy rats, those treated with clofibric acid for 12 weeks had lower plasma cholesterol and markedly less albuminuria and FGS.[13] Tubulointerstitial disease and vascular lesions, which are also characteristic of renal disease in $\frac{5}{6}$ nephrectomy rats, were both reduced in clofibric acid-treated rats. Renal function, assessed by serum creatinine levels and inulin clearance, was preserved in clofibric acid-treated $\frac{5}{6}$ nephrectomy rats. The beneficial effects of clofibric acid on renal structure and function in $\frac{5}{6}$ nephrectomy rats were realized without an effect of the drug on the elevated blood pressure that occurs in this model of chronic renal failure.

Reduction of renal disease with clofibric acid was also demonstrated in obese Zucker rats.[19] These rats have been studied as an experimental model of non-insulin-dependent diabetes mellitus (NIDDM).[51] Obese Zucker rats demonstrate insulin resistance, hyperinsulinemia, hyperlipidemia, and glucose intolerance that yields to mild fasting hyperglycemia with age. Importantly, obese Zucker rats develop progressive renal disease that has been studied as a model of nephropathy in NIDDM.[51] In this regard, mesangial matrix expansion and "microalbuminuria" are present in young (9 to 14 weeks of age) obese Zucker rats.[52,53] Severe albuminuria, FGS, and reduced GFR become evident by 28 weeks of age, and renal disease is progressive.[52] In a recent study, clofibric acid, administered from 8 weeks of age, reduced mesangial matrix expansion, mesangial cellularity, and the incidence of FGS in 40-week-old obese rats.[19] Unfortunately, no assessment of renal function was performed in that study to determine whether longer-term clofibric acid also preserved GFR in obese Zucker rats.

The mechanism by which fibrates such as halofenate and clofibric acid ameliorate experimental renal injury is unknown. Clofibric acid inhibits hepatic cholesterol synthesis and lipoprotein release, and stimulates lipoprotein lipase activity.[54] Recent studies have demonstrated lipoprotein binding and uptake by glomerular cells.[55,56] Moreover, there is evidence that lipoproteins can directly injure glomerular cells.[57–59] Thus, reduction of circulating lipoproteins by clofibric acid may have had direct benefits by reducing the levels of potentially cytotoxic lipoproteins in the glomerulus. It is unlikely that the fibrates exerted their beneficial effects through favorable actions on glomerular hemodynamics, such as a reduction in glomerular pressures. Micropuncture studies demonstrated that single nephron function and glomerular ultrafiltration dynamics were similar in untreated obese Zucker rats and obese rats treated for 6 weeks with clofibric acid.[53]

Several clinical studies have examined the effects of fibrates in treating hyperlipidemia secondary to the nephrotic syndrome. Clofibrate has been used in such studies, but, because of complications of muscle injury, has been generally avoided. A recent study examined the treatment of nephrotic hyperlipidemia with an alternate agent, gemfibrozil.[60] Nephrotic patients were randomized to a double-blind, placebo-controlled crossover study of gemfibrozil for two 6-week periods. During the gemfibrozil period, plasma triglycerides fell by 50 percent, while reductions in total and LDL cholesterol were more modest. There were no effects of gemfibrozil on either creatinine clearance or urine protein excretion. Whether for a longer term treatment of nephrotic syndrome with agents such as gemfibrozil can affect the disease course remains to be determined.

HMG-CoA Reductase Inhibitors

The enzyme 3-hydroxy-3-methylglutaryl-coenzyme A (HMG-CoA) reductase is present in most mammalian cells and catalyzes the rate-limiting step in cholesterol biosynthesis. Several HMG-CoA reductase inhibitors have been developed, and their effects on circulating lipids and renal disease have been experimentally and clinically evaluated.

Several studies have examined the effects of lovastatin (formerly mevinolin) on the course of experimental renal disease. In $\frac{5}{6}$ nephrectomy rats and obese Zucker rats, lovastatin, like clofibric acid, reduced serum cholesterol and dramatically reduced the extent of renal injury.[13,19] Lovastatin also reduced serum triglycerides in both experimental models.[13,19] In $\frac{5}{6}$ nephrectomy rats, lovastatin administration for 10 weeks reduced albuminuria, mesangial matrix expansion, and tubulointerstitial injury.[13] FGS was reduced by more than 50 percent in the lovastatin-treated rats.[13] Lovastatin had no effect on elevated blood pressure in $\frac{5}{6}$ nephrectomy rats.[13] No measurements of either serum creatinine or inulin clearance in lovastatin-treated $\frac{5}{6}$ nephrectomy rats were performed, however, to determine whether GFR was preserved.

Lovastatin had similar effects on glomerular disease in obese Zucker rats. Obese rats treated with daily lovastatin for 32 weeks had an approximately 75 percent reduction in albuminuria and significantly less mesangial matrix and cellularity compared with untreated obese Zucker rats.[19] Moreover, the incidence of FGS in the lovastatin-treated obese rats was only approximately 10 percent of that in the untreated obese rats.[19] It was not determined whether lovastatin treatment preserved GFR in obese Zucker rats. Micropuncture studies in obese rats, however, did demonstrate no significant effects of short-term (6 weeks) lovastatin treatment on single nephron function and the determinants of glomerular ultrafiltration.[19]

Recently, the effects of lovastatin on glomerular injury in experimental nephrotic syndrome were examined.[28] PAN rats were treated for 60 days with lovastatin and compared with untreated PAN rats. Lovastatin reduced plasma cholesterol and triglycerides in PAN rats by approximately 40 to 50 percent, but had no effect on albuminuria. Importantly, however, renal lesions at 60 days were markedly less in the lovastatin-treated nephrotic rats. Fewer proteinaceous casts and dilated tubules, and less glomerulosclerosis and interstitial mononuclear cell infiltration were observed in lovastatin-treated PAN rats. Moreover, at 60 days, inulin clearance in the lovastatin-treated nephrotic rats was nearly twice that in the untreated nephrotic rats. Thus, lovastatin significantly preserved both renal structure and function in rats with experimental nephrotic syndrome.

Preliminary studies have recently demonstrated that lovastatin also reduces glomerular injury in Dahl S rats.[30] Dahl S rats have frequently been studied as one model of human essential hypertension. These rats develop systemic hypertension when fed a high-salt diet.[29,31,32] Hyperlipidemia and progressive renal disease are also characteristic of Dahl S rats,[29,31,32] and it has been suggested that lipids may be responsible, at least in part, for the initiation of glomerular injury in this model.[29] Dahl S rats treated with daily lovastatin for 22 weeks had normal plasma cholesterol, and had markedly less albuminuria and FGS than untreated Dahl S rats.[30] Interestingly, blood pressure in Dahl S rats was also reduced by lovastatin.[30]

Another recent preliminary study investigated the effects of pravastatin on serum lipids and urine protein excretion in streptozotocin diabetic rats.[61] Rats were treated for 3 months with the drug administered via the drinking water. Interestingly, pravastatin did not lower the elevated serum lipids in the diabetic rats, but did significantly reduce proteinuria. Moreover, pravastatin reduced glomerular basement membrane thickening. Thus, pravastatin appeared to exert a beneficial effect in experimental diabetic nephropathy independent of an action on total circulating lipid levels. Whether pravastatin affected the lipid composition of circulating lipoproteins, however, is unknown.

It is important to note that in all of the above experimental studies, the HMG-CoA reductase inhibitors were administered preventatively. That is, administration of the cholesterol synthesis inhibitors was initiated at a very young age or at the time when a particular maneuver (e.g., $\frac{5}{6}$ nephrectomy, PAN injection) designed to induce renal disease was performed. The fact that such a pharmacologic regimen uniformly reduced renal disease in the various experimental models has provided strong evidence that lipids are important in at least the initiation of glomerular injury. It remains to be determined, however, whether intervention therapy with hypolipidemic drugs can slow the progression of established renal disease. Such studies are critical to understanding the importance of lipids in the progression of renal disease. These studies are also crucial to defining whether, in the clinical setting, hypolipidemic agents might possibly slow the rate of functional decline and the progression of structural injury in patients with established renal disease.

A recent preliminary study investigated the effects of lovastatin on the progression of established glomerular disease in obese Zucker rats.[62] Lovastatin was administered to obese Zucker rats beginning at 26 weeks of age (i.e., when marked albuminuria was evident). At 44 weeks, lovastatin-treated obese rats had a 50 percent reduction in the incidence of FGS compared with vehicle-treated, control obese rats. Thus, lovastatin significantly arrested the course of established glomerular disease in this model, suggesting that lipids play an important role in the progression of renal disease.

A recent clinical study examined the effects of chronic (48 weeks) simvastatin therapy on serum lipids and renal function in seven patients with nephrotic syndrome.[63] Total and LDL cholesterol were reduced by simvastatin throughout the treatment period. Importantly, urine albumin excretion declined significantly (by 60 percent) and plasma protein concentration rose concomitantly. Serum creatinine remained stable throughout the study period. Thus, this study suggests that prolonged pharmacologic reduction of serum lipids may have beneficial effects in retarding the progression of clinical renal disease. It must be noted, however, that this was an uncontrolled study, that measurements of GFR were not reported, and that there was no information regarding renal morphology.

Bile Acid-Binding Resins

Compounds such as cholestyramine and colestipol are sequestering agents that partially block the enterohepatic circulation of bile acids. These bile acid-sequestering agents have been shown, experimentally and clinically, to lower elevated plasma cholesterol levels. In nephrotic rats, cholestyramine administered in the diet significantly blunted initial increases in urine protein excretion, plasma cholesterol, and triglycerides 1 to 2 weeks after an injection of PAN.[64] Moreover, PAN rats treated chronically with cholestyramine for 24 weeks had much less glomerular injury compared with untreated nephrotic rats.[64] Cholestyramine-treated PAN rats had less urine protein excretion during the phase of recurrent proteinuria (15 to 24 weeks) in this model, and histopathologic injury was less. The percentage of glomeruli with glomerulosclerosis and/or hyalinosis in the cholestyramine-treated PAN rats was less than one-third that in the untreated nephrotic rats. As a functional correlate, GFR at 24 weeks was relatively preserved in the cholestyramine-treated PAN rats, while untreated nephrotic rats demonstrated significant renal insufficiency. Thus, this particular bile acid-sequestering agent was effective in ameliorating the glomerular structural injury and renal insufficiency that characterize experimental nephrotic syndrome.

Both cholestyramine and colestipol have been used clinically to treat hyperlipidemia associated with the nephrotic syndrome, and both agents have successfully reduced total and LDL cholesterol levels in nephrotic patients.[65–67] In one recent study, cholestyramine treatment for 6 weeks in a small group of nephrotic patients had no effect on either serum creatinine or albuminuria.[67] The effects of longer-term cholestyramine treatment on renal function and albuminuria in patients with established renal disease have not been investigated.

Probucol

Probucol is a relatively new hypolipidemic agent that has been shown to be very effective in the treatment of primary and familial hypercholesterolemia. Probucol is also an antioxidant. Only a few reports are presently available with respect to probucol effects in experimental renal disease. Blantz and co-workers recently found that probucol did not block the reductions in GFR, single nephron glomerular filtration rate (SNGFR), single nephron plasma flow (SNPF), and the glomerular ultrafiltration coefficient that occur acutely in rats treated with antiglomerular basement membrane antibody.[68] Moreover, probucol did not reduce glomerular neutrophil infiltration in this model of glomerular immune injury. The effects of probucol on serum lipids and proteinuria in anti-glomerular basement membrane rats were not reported.

In another study, Kaplan and co-workers found that probucol prevented decreases in SNGFR and SNPF, and lowered glomerular capillary pressure and renal arteriolar resistance in rats fed a high-cholesterol diet.[69] Interestingly, the actions of probucol on glomerular hemodynamics occurred in the absence of an effect of probucol on total serum cholesterol levels. Elevated urine thromboxane in cholesterol-fed rats was also prevented by probucol treatment. It was suggested, therefore, that oxidized lipoproteins in cholesterol-fed rats caused renal vasoconstriction via enhanced thromboxane production, and that this was blocked by the antioxidant activity of probucol. The effects of probucol on albuminuria and glomerular structural injury that develop with time in cholesterol-fed rats (see next section) have not been reported.

In streptozotocin-diabetic rats, probucol treatment for 3 months lowered serum cholesterol and triglycerides, but had no effects on glomerular basement membrane thickening or urine albumin excretion.[61] Thus, the effects of probucol in experimental models of renal disease appear to be variable. However, systematic studies of the effects of chronic probucol treatment on glomerular hemodynamics and glomerular histologic injury in various models of experimental renal disease remain to be performed.

Probucol has also been used clinically to treat hyperlipidemia in nephrotic patients. In one recent study, probucol treatment for 12 weeks in a group of nephrotic patients lowered serum total cholesterol, LDL cholesterol, and triglycerides.[70] Creatinine clearance and urine protein excretion in the nephrotic patients were unaffected by probucol therapy. The effects of longer-term probucol treatment on proteinuria and progression of renal disease have not been reported.

DIETARY MANIPULATION OF PLASMA LIPIDS: EFFECTS ON RENAL STRUCTURE AND FUNCTION

Diet-Induced Hyperlipidemia

In 1967, French et al. reported that dietary cholesterol supplementation caused glomerular injury in guinea pigs.[42] Glomerular histopathology in the cholesterol-fed animals was characterized by an increase in mesangial periodic acid-Schiff–positive material, as well as by abundant lipid deposition. Subsequent studies have further examined the effects of dietary cholesterol and/or saturated fat supplementation on the development of glomerular injury in several species. The effects of diet-induced hypercholesterolemia on the glomerulus are summarized in Table 7-2.

Table 7-2. Effects of Diet-Induced Hypercholesterolemia on Renal Structure and Function

	Effect
Kidney weight	Normal
Glomerular area	↑
Mesangial matrix, fractional mesangial area	↑
Mesangial cellularity[a]	↑
Lipid deposits (oil red O)	↑
Epithelial cell protein droplets	↑
Proteinuria/albuminuria	↑
Tubulointerstitial injury	↑
Glomerulosclerosis	↑
P_{GC}	↑
SNGFR	Normal or ↓
SNPF	Normal or ↓

Abbreviations: P_{GC}, glomerular capillary pressure; SNGFR, single nephron glomerular filtration rate; SNPF, single nephron plasma flow.

[a] Includes cells exhibiting monocyte/macrophage characteristics.

In another study in guinea pigs, Al-Shebeb and colleagues documented serial glomerular changes that occurred in response to high cholesterol intake.[43] Serum cholesterol was elevated fourfold at day 5, and rose progressively throughout the experimental period. Oil red O-positive lipid deposits were detected in glomeruli after only 5 days of high cholesterol intake, and were extensive at day 70. At 30 days, significant mesangial expansion was present in the cholesterol-fed animals, and this was progressive over the next 40 days. Proteinuria and increased glomerular monocytes were evident at 70 days. Renal function, assessed by creatinine clearance, was not altered over the experimental period. Thus, high cholesterol intake in guinea pigs caused progressive glomerular structural injury that was preceded by the development of hypercholesterolemia and glomerular lipid deposition, and was accompanied by glomerular accumulation of monocytes.

Several studies have demonstrated accelerated glomerular injury in rats fed diets rich in cholesterol with or without saturated fat.[44–46] Gröne et al. recently showed that rats fed a diet high in cholesterol and saturated fat had increased urine protein excretion at 80 days.[45] After 5 to 6 months on the same diet, rats demonstrated significant mesangial expansion and more than a sevenfold increase in the percentage of glomeruli with sclerosis compared with age-matched control rats fed a standard diet. No direct measurements of renal function were performed in this study. Serum creatinine at 150 to 180 days, however, was not altered in the rats fed the high-cholesterol, high-fat diet.

Kasiske and co-workers reported similar results in rats fed a 4 percent cholesterol diet.[46] After 4 weeks, serum cholesterol was increased more than twofold, and fractional mesangial area was increased by 29 percent in the cholesterol-fed rats compared with control rats fed standard chow. Interestingly, glomerular area in cholesterol-fed rats was increased 17 percent at 4 weeks. After 19 weeks, cholesterol-fed rats had marked albuminuria and a fourfold increase in the incidence of FGS. Increased mesangial matrix and cellularity were also observed in the cholesterol-fed rats. The increased cellularity appeared to be due, at least in part, to an increase in glomerular macrophages. Use of monoclonal antibodies directed against rat lymphohemopoietic cell surface markers showed an increase in glomerular cells exhibiting macrophage surface markers in rats fed the high-cholesterol diet. These results were similar to those in guinea pigs fed a high-cholesterol diet, in which increased glomerular monocyte counts were demonstrated.[43]

Kasiske et al. also performed micropuncture studies in separate groups of cholesterol-fed rats and control rats. After 8 to 10 weeks, cholesterol-fed rats had elevated glomerular pressures compared with control rats fed standard chow.[46] Plasma colloid osmotic pressure was also elevated in the cholesterol-fed rats, and correlated significantly with tubular stop-flow pressure. There were no differences in either SNGFR or SNPF between cholesterol- and standard chow-fed rats. A trend toward elevated efferent arteriolar resistance was observed in the cholesterol-fed rats. These results were similar in some respects to those reported by Kaplan et al. in cholesterol-fed rats.[69]

Thus, it appears that diet-induced hypercholesterolemia in rats can cause specific alterations in glomerular hemodynamics. Importantly, the rise in glomerular pressures could contribute, at least in part, to the pathogenesis of glomerular injury in this model of renal disease. Interestingly, in a preliminary study, elevated glomerular pressures were found in 24- to 26-week-old obese Zucker rats, at an age when these rats were developing severe hypercholesterolemia and albuminuria.[71] These results are consistent with the micropuncture studies in cholesterol-fed rats[46,69] and suggest that hypercholesterolemia per se may cause an elevation of glomerular pressures. The mechanism of such a lipid-hemodynamic interaction in the glomerulus remains to be determined, but it has been suggested that oxidized lipoproteins may alter glomerular hemodynamics by stimulating thromboxane production.[69]

Diet-induced hypercholesterolemia has also been shown to exaggerate glomerular disease induced by other maneuvers or by immune injury. Nephrotic rats fed a high-cholesterol or high-fat diet had more extensive glomerular injury than nephrotic rats fed standard chow.[72] Diet-induced hypercholesterolemia also exaggerated glomerular disease in the unclipped kidneys of rats with two-kidney, one-clip hypertension.[45] Nephrectomized rats fed a high-cholesterol diet had higher glomerular pressures and approximately four times more FGS than nephrectomized rats fed standard chow.[46] Lastly, a diet high in saturated fat caused elevated serum cholesterol levels and exaggerated the extent of glomerular immune injury in NZB/NZW mice.[73]

Polyunsaturated Fatty Acid Supplementation and Renal Disease

Dietary enrichment with polyunsaturated fatty acids (PUFAs) as a means of altering the progression of glomerular injury has received considerable attention in the past decade. PUFAs with carbon-carbon double bonds in the sixth (n-6) and third (n-3) positions are considered essential fatty acids. The n-6 class of PUFAs is primarily derived from the oil of plant seeds. One of these fatty acids, linoleic acid, is a precursor for the synthesis of arachidonic acid and, accordingly, the dienoic eicosanoids such as prostaglandin E_2 (PGE_2), thromboxane A_2 (TXA_2), and prostacyclin (PGI_2). In contrast, n-3 PUFAs are primarily derived from the oil of marine vertebrates. One of these fatty acids, eicosapentaenoic acid (EPA), is the precursor for the synthesis of trienoic eicosanoids such as PGE_3, TXA_3, and PGI_3. The effects of dietary PUFA supplementation in some experimental models of renal disease are summarized in Table 7-3.

Hurd et al. initially investigated whether changes in the fatty acid composition of the diet might influence the course of experimental renal disease.[74] These investigators studied NZB/NZW mice, a strain characterized by an autoimmune glomerulonephritis analogous to that which is seen in clinical

Table 7-3. Effects of Dietary Supplementation with n-6 or n-3 Polyunsaturated Fatty Acids on Renal Structure and Function

	Glomerular Injury		Renal Function	
	n-6	n-3	n-6	n-3
Murine lupus nephritis	↔	↓	NR	NR
Apoferritin glomerulonephritis (mice)	↓	↓	NR	NR
Subtotal nephrectomy (rat)	↓	↑ or ↓	P	↓ or P
Normal rats	—	—	N	N or ↑
Humans	—	—	↑	↑

Abbreviations: NR, not reported; N, normal; P, preserved; ↔, no effect.

lupus nephritis. NZB/NZW mice develop circulating autoantibodies, glomerular immune complex deposits, glomerular hypercellularity, proteinuria, and glomerulosclerosis, and ultimately die from terminal renal failure.

Interestingly, a diet high in safflower oil (78 percent linoleic acid) did not reduce mortality or ameliorate glomerular injury in NZB/NZW mice.[74] On the other hand, NZB/NZW mice fed an essential fatty acid-deficient (EFAD) diet had markedly improved survival and less proteinuria and glomerular histologic injury. The precise mechanism of the beneficial effects of the EFAD diet in this model is unknown, but recent studies have indicated that an EFAD diet can reduce the number of Ia^+ cells (macrophages) in the glomerular mesangium.[75] Thus, it is conceivable that the EFAD diet reduced glomerular disease in NZB/NZW mice by an action on glomerular macrophages.

Kelley and co-workers also were unable to demonstrate a beneficial effect of dietary safflower oil supplementation on longevity or glomerular injury in MRL/lpr mice, a more severe murine model of lupus nephritis.[76] In contrast, dietary supplementation with fish oil (n-3 PUFAs) improved survival and decreased proteinuria in MRL/lpr mice.[76] Moreover, MRL/lpr mice fed fish oil had less glomerular lymphocyte proliferation, macrophage surface Ia expression, and interstitial inflammation than control mice fed a standard diet.[76]

Similar results with n-3 PUFAs were obtained in two studies of NZB/NZW mice. Diets containing high levels of n-3 PUFAs improved survival and markedly reduced proteinuria and glomerular injury in NZB/NZW mice.[77,78] Thus, in two murine models of lupus nephritis, dietary supplementation with n-3 PUFAs reduced immune-mediated glomerular disease. Dietary n-6 PUFA supplementation, on the other hand, had no effects on the course of glomerular disease.

Beneficial effects of dietary PUFA supplementation have also been reported in a murine model of immune complex glomerulonephritis produced by repeated intraperitoneal injections of apoferritin. Interestingly, Kher et al. demonstrated a protective effect of both n-6 and n-3 PUFAs on the development of glomerular disease in apoferritin-injected mice.[79] Specifically, diets supplemented with either linoleic acid or eicosapentaenoic acid reduced proteinuria and glomerular histologic changes in this model.

The mechanism of the beneficial effects of PUFAs, particularly n-3 PUFAs, in immune complex glomerulonephritis is unclear. In some of the above studies, changes in eicosanoid production in PUFA-supplemented animals were quantitated. In general, an increase in tissue and/or urine dienoic prostaglandins and prostaglandin metabolites occurred in response to dietary n-6 PUFA supplementation, and a decrease in these substances occurred with n-3 PUFA supplementation. The latter phenomenon would be expected since n-3 and n-6 PUFAs compete for the cyclooxygenase enzymes responsible for eicosanoid biosynthesis.

Dietary supplementation with n-3 PUFAs has been shown to increase renal production of trienoic eicosanoids.[76,79] It is possible that such alterations may underlie, in part, the beneficial effects of n-3 PUFAs in immune-mediated glomerular injury. Indeed, because eicosanoids can affect cellular growth and proliferation,[80–82] it is tempting to speculate that dietary n-3 PUFAs may have important consequences in regulating these components of progressive renal disease. Dietary fish oil supplementation also has been recently shown to inhibit the release of growth factors, tumor necrosis factor, and interleukin-1 from different cell types, including macrophages.[83,84] Finally, dietary PUFAs may cause physical alterations in cell membranes, leading to enhanced fluidity and altered membrane function.[85]

Of important clinical relevance, a preliminary report has indicated that dietary supplementation with n-3 PUFAs may be beneficial in altering the course of human immunoglobulin A nephropathy.[86] A group of 11 patients with IgA nephropathy were given daily supplements of eicosapentaenoic acid for 1 year. At the end of this period, a significant reduction in proteinuria and significantly greater GFR (iothalamate clearance) were observed. Moreover, hypertensive and normotensive patients improved equally. In contrast to this report, however, Cheng and colleagues found no beneficial effects of short-term fish oil supplementation on renal function or proteinuria in patients with IgA nephropathy.[87] Thus, the effects of n-3 PUFA supplementation on clinical renal disease are unclear and await further systematic study.

Dietary Polyunsaturated Fatty Acid Supplementation: Effects on Experimental Chronic Renal Failure

Recent studies have examined the effects of dietary enrichment with linoleic acid in subtotally nephrectomized rats. In $\frac{3}{4}$ nephrectomized rats, Barcelli and colleagues demonstrated a significant reduction in the severity of glomerular and tubular lesions and the degree of proteinuria after 20 weeks of dietary linoleic acid supplementation.[88] Increased glomerular areas and elevated systemic blood pressure in the nephrectomized rats, however, were unaltered by linoleic acid supplementation. The beneficial actions of dietary linoleic acid on glomerular disease in the subtotal nephrectomy model have been confirmed in subsequent studies.[89,90] These studies consistently demonstrated significant reductions in glomerulosclerosis and proteinuria with linoleic acid supplementation. Moreover, renal function in nephrectomized rats was preserved with n-6 supplementation, and increases in urinary and tissue dienoic eicosanoids were demonstrated.[88–90]

The effects of dietary n-3 PUFA supplementation on glomerular injury in the subtotal nephrectomy model of chronic renal failure are controversial. Barcelli and co-workers demonstrated a beneficial effect of dietary salmon oil supplementation in reducing glomerular injury and preserving renal function in $\frac{3}{4}$ nephrectomized rats.[91] In another study, Clark et al. demonstrated a reduction in albuminuria in $\frac{5}{6}$ nephrectomy rats fed fish oil for 28 days.[92] In contrast to these findings, Scharschmidt and colleagues demonstrated a deleterious effect of dietary n-3 PUFA enrichment in subtotally nephrectomized rats.[93] Rats with either 1-$\frac{1}{3}$ or 1-$\frac{2}{3}$ nephrectomy were fed a diet containing approximately 20 percent menhaden oil. In one study, 1-$\frac{2}{3}$ nephrectomized rats fed menhaden oil had significantly increased mortality. In a separate study, 1-$\frac{1}{3}$ nephrectomized rats fed fish oil for 12 weeks had more proteinuria and glomerulosclerosis than nephrectomized rats fed standard chow. Thus, in these studies, dietary fish oil supplementation exaggerated glomerular disease. This occurred despite a reduction in plasma triglycerides and a modest decrease in plasma cholesterol. In short, the effects of dietary n-3 PUFA supplementation in the subtotal nephrectomy model of chronic renal failure are controversial, and a resolution of the differences in the results of the above studies awaits further investigation.

Recently, the effects of dietary PUFA supplementation on glomerular hemodynamics have been examined in rats with normal and reduced renal mass. Such studies are important for understanding the overall effects of dietary PUFA supplementation on the course of glomerular disease, since alterations in glomerular hemodynamics have been described as risk factors for the development of glomerulosclerosis. In particular, elevations in glomerular pressures have been implicated in the pathogenesis of FGS.[94,95] In a preliminary study, Schmitz and co-workers found that Sprague-Dawley rats fed a diet containing 20 percent menhaden oil had a marked increase in SNGFR.[96] This was due primarily to increased SNPF that resulted from a significant 37 percent decrease in efferent arteriolar resistance. Importantly, despite increases in SNGFR and SNPF, glomerular pressures were not altered in the rats fed fish oil. Thus, dietary supplementation with n-3 PUFAs caused glomerular hyperfiltration and hyperperfusion, but not glomerular hypertension. Dietary supplementation with 20 percent sunflower oil (n-6 PUFAs), in contrast, had no effect on glomerular hemodynamics.[96]

Logan and co-workers did not find a change in GFR, normalized for body weight, in two-kidney rats fed a fish oil-supplemented diet.[97] In uninephrectomized rats, however, dietary fish oil supplementation caused increases in both GFR and RPF. Thus, there is evidence that dietary n-3 PUFAs can cause renal vasodilatation and increases in nephron plasma flow and filtration rate. Whether similar PUFA-mediated functional changes occur in a setting of renal disease and influence the rate of disease progression remains to be determined.

Dietary Polyunsaturated Fatty Acid Supplementation and Renal Function in Humans

Recent studies have indicated that dietary PUFA supplementation can also alter renal function in humans. In normal males and females given varying concentrations of linoleic acid for 2 weeks, the renal excretion of creatinine and sodium increased in proportion to the percent of linoleic acid in the diet.[98] Bernardi et al. recently administered intravenous linoleate (1.24 mg/kg) to healthy subjects and patients with chronic renal failure, liver cirrhosis, or congestive heart failure.[99] Measurements of renal function were obtained at 30, 60, and 90 minutes after the linoleate bolus. Significant increments in creatinine clearance, para-aminohippurate clearance, and urinary sodium excretion occurred in all cases. In both of the above studies, a significant rise in urine dienoic prostaglandin excretion also occurred in response to linoleic acid administration. Furthermore, the renal hemodynamic and natriuretic effects of linoleic acid were abolished by administration of a prostaglandin synthetase inhibitor.

Similar changes in renal function have been found in response to dietary n-3 PUFA supplementation. Düsing et al. recently demonstrated significant increases in inulin clearance and para-aminohippurate clearance in patients given 6 weeks of dietary n-3 PUFA supplements.[100] Moreover, filtration fraction fell, suggesting a preferential reduction in efferent arteriolar resistance. These results were consistent with the glomerular hemodynamic changes reported by Schmitz et al. in normal rats fed a fish oil-supplemented diet.[96] There was little change in blood pressure or urinary excretion of sodium, potassium, or chloride in patients given n-3 PUFA supplementation.[100] Urinary excretion of PGE_2 fell with dietary n-3 PUFA enrichment, and there was a reciprocal increase in the urinary excretion of PGE_3.[100] Lastly, a diet high in n-3 PUFAs has been shown to augment GFR and renal blood flow in patients with chronic renal failure.[101] Whether these functional effects of n-3 PUFAs might influence the course of glomerular injury in these patients is unknown and is an important area of future investigation.

REFERENCES

1. Virchow R: A more precise account of fatty metamorphosis. p. 342. In Chance (ed): Cellular Pathology. 2nd Ed. Gryphon Editors, Birmingham, 1860
2. Munk F: Beriche über Krankheitsfälle und Behandlungsverfahren. Med Klin 12:1019, 1916

3. Fahr T: Beträge zur Frage der Nephrose. Virchows Arch [A] 239:32, 1922
4. Bell ET: Lipoid nephrosis. Am J Pathol 5:587, 1929
5. Kimmelstiel P, Wilson C: Intercapillary lesions in the glomerulus of the kidney. Am J Pathol 12:83, 1936
6. Newburger RA, Peters JP: Intercapillary glomerulosclerosis. A syndrome of diabetes, hypertension, and albuminuria. Arch Intern Med 64:1252, 1939
7. Wilens SL, Elster SK: The role of lipid deposition in renal arteriolar sclerosis. Am J Med Sci 219:183, 1951
8. Keane WF, Kasiske BL, O'Donnell MP: Lipids and progressive glomerulosclerosis. A model analogous to atherosclerosis. Am J Nephrol 8:261, 1988
9. Moorhead JF, Chan MK, El-Nahas M et al: Lipid nephrotoxicity in chronic progressive glomerular and tubulo-interstitial disease. Lancet 1:1309, 1982.
10. Diamond JR, Karnovsky MJ: Focal and segmental glomerulosclerosis. Analogies to atherosclerosis. Kidney Int 33:917, 1988
11. Edwards KDG: Why treat hyperlipidemia in patients with renal disease? p. 3. In: Controversies in Nephrology. Vol. 3. Nephrology Division, Georgetown University, Georgetown, WA 1981
12. Schmitz PG, Kasiske BL, O'Donnell MP, Keane WF: Lipids and progressive renal injury. Semin Nephrol 9:354, 1989
13. Kasiske BL, O'Donnell MP, Garvis WJ, Keane WF: Pharmacological treatment of hyperlipidemia reduces glomerular injury in rat $\frac{5}{6}$ nephrectomy model of chronic renal failure. Circ Res 62:367, 1988
14. Chobanian AV, Arquilla ER, Clarkson TB et al: Cardiovascular complications. Diabetes, suppl. 1 31:54, 1982
15. Michels LD, O'Donnell MP, Keane WF: Glomerular hemodynamic and structural correlations in long-term experimental diabetic rats. J Lab Clin Med 103:840, 1984
16. Zatz R, Meyer TW, Rennke HG, Brenner BM: Predominance of hemodynamic rather than metabolic factors in the pathogenesis of diabetic glomerulopathy. Proc Natl Acad Sci USA 82:5963, 1985
17. Velasquez MT, Kimmel PL, Michaelis OE: Animal models of spontaneous diabetic kidney disease. FASEB J 4:2850, 1990
18. Kasiske BL, Cleary MP, O'Donnell MP, Keane WF: Effects of genetic obesity on renal structure and function in the Zucker rat. J Lab Clin Med 106:598, 1985
19. Kasiske BL, O'Donnell MP, Cleary MP, Keane WF: Treatment of hyperlipidemia reduces glomerular injury in obese Zucker rats. Kidney Int 33:667, 1988
20. Koletsky S: Pathologic findings and laboratory data in a new strain of obese hypertensive rats. Am J Pathol 80:129, 1975
21. Velasquez MT, Kimmel PL, Michaelis OE et al: Effect of carbohydrate intake on kidney structure and function in SHR/N-cp rats. A new model of NIDDM. Diabetes 38:679, 1989
22. Velasquez MT, Michaelis OE, Szallasi T et al: Glomerular hypertrophy and mesangial expansion in the SHR/N-cp rat with type II diabetes: role of type of carbohydrate diet, abstracted. Kidney Int 37:523, 1990
23. Gartner K: Glomerular hyperfiltration during the onset of diabetes mellitus in two strains of diabetic mice (C57BL/6J db/db and C57BL/KsJ db/db). Diabetologia 15:59, 1978
24. Bower G, Brown DM, Steffes MW et al: Studies on the glomerular mesangium and the juxtaglomerular apparatus in the genetically diabetic mouse. Lab Invest 43:333, 1980

25. Grond J, Weening JJ, Elema JD: Glomerular sclerosis in nephrotic rats. Comparison of the long-term effects of Adriamycin and aminonucleoside. Lab Invest 51:277, 1984
26. Whiteside C, Prutis K, Cameron R, Thompson J: Glomerular epithelial cell detachment, not reduced charge density, correlates with proteinuria in Adriamycin and puromycin nephrosis. Lab Invest 61:650, 1989
27. Scholey JW, Miller PL, Rennke HG, Meyer TW: Effect of converting enzyme inhibition on the course of Adriamycin-induced nephropathy. Kidney Int 36:816, 1989
28. Harris KPG, Purkerson ML, Yates J, Klahr S: Lovastatin ameliorates the development of glomerulosclerosis and uremia in experimental nephrotic syndrome. Am J Kidney Dis 15:16, 1990
29. O'Donnell MP, Kasiske BL, Keane WF: Risk factors for glomerular injury in rats with genetic hypertension. Am J Hypertens 2:9, 1989
30. O'Donnell MP, Kasiske BL, Schmitz PG et al: Contrasting effects of lovastatin and enalapril on glomerular injury in Dahl salt-sensitive rats, abstracted. Kidney Int 37:391, 1990
31. Sterzel RB, Luft FC, Gao Y et al: Renal disease and the development of hypertension in salt-sensitive Dahl rats. Kidney Int 33:1119, 1988
32. Raij L, Azar S, Keane WF: Role of hypertension in progressive glomerular immune injury. Hypertension 7:398, 1985
33. Mondon CE, Reaven GM: Evidence of abnormalities of insulin metabolism in rats with spontaneous hypertension. Metabolism 37:303, 1988
34. Mondon CE, Reaven GM, Azhar S et al: Abnormal insulin metabolism by specific organs from rats with spontaneous hypertension. Am J Physiol 257:E491, 1989
35. Imai Y, Matsumura H, Miyajima H, Oka K: Serum and tissue lipids and glomerulonephritis in the spontaneously hypercholesterolemic rats, with a note on the effects of gonadectomy. Atherosclerosis 27:165, 1977
36. Brown DM, Steffes MW, Thibert P, Azar S, Mauer SM: Glomerular manifestations of diabetes in the BB rat. Metabolism suppl. 1 32:131, 1983
37. Cohen AJ, McGill PD, Rossetti RG et al: Glomerulopathy in spontaneously diabetic rat. Diabetes 36:944, 1987
38. Zatz R, Prado EBA, Prado MJBA et al: Aging analbuminemic rats fail to develop focal glomerular sclerosis (FGS), abstracted. Kidney Int 35:442, 1989
39. Raij L, Tolins JP, Luscher T: Hyperlipemia and renal injury. Studies in atherosclerosis prone Watanabe rabbits with hereditary hyperlipemia, abstracted. Kidney Int 33:383, 1988
40. Kasiske BL, O'Donnell MP, Schmitz PG, Keane WF: The role of lipid abnormalities in the pathogenesis of chronic, progressive renal disease. p. 109. In Grünfeld J-P, Bach JF, Funck-Bretano J-L, Maxwell MH (eds): Advances in Nephrology. Vol. 20. Mosby-Year Book Medical Publishers, Chicago, 1991
41. Wellmann KF, Volk BW: Renal lesions in experimental hypercholesterolemia in normal and sub-diabetic rabbits. II. Longterm studies. Lab Invest 24:144, 1971
42. French SW, Yamanaka W, Ostwald R: Dietary induced glomerulosclerosis in the guinea pig. Arch Pathol Lab Med 83:204, 1967
43. Al-Shebeb T, Frohlich J, Magil AB: Glomerular disease in hypercholesterolemic guinea pigs: a pathogenetic study. Kidney Int 33:498, 1988
44. Peric-Golia L, Peric-Golia M: Aortic and renal lesions in hypercholesterolemic adult, male, virgin Sprague-Dawley rats. Atherosclerosis 46:57, 1983

45. Gröne HJ, Walli A, Gröne E et al: Induction of glomerular sclerosis by a lipid-rich diet in male rats. Lab Invest 60:433, 1989
46. Kasiske BL, O'Donnell MP, Schmitz PG, Keane WF: The renal injury of diet-induced hypercholesterolemia in rats. Kidney Int 37:880, 1990
47. Amaturda JM, Margolus S, Hutchins GM: Type III hyperlipoproteinemia with mesangial foam cells in renal glomeruli. Arch Pathol Lab Med 98:51, 1974
48. Saito T, Sato H, Kudo K et al: Lipoprotein glomerulopathy: glomerular lipoprotein thrombi in a patient with hyperlipoproteinemia. Am J Kidney Dis 13:148, 1989
49. Edwards KDG, Strachan JC, Levy SA, Chui C-F: Protection by L-trytophan or halofenate against hyperlipidemia and acute or chronic aminonucleoside-nephrosis in sucrose/lard-fed rats, abstracted. Kidney Int 12:480, 1977
50. Edwards KDG, Strachan JC: Aminonucleoside-induced chronic renal failure in rats: enhancement by dietary hyperlipidemias and protection by a hypolipidemic diet or halofenate, abstracted. Kidney Int 14:723, 1978
51. Kasiske BL, O'Donnell MP, Keane WF: The obese Zucker rat model of glomerular injury in type II diabetes. J Diabetic Complications 1:26, 1987
52. Kasiske BL, Cleary MP, O'Donnell MP, Keane WF: Effects of genetic obesity on renal structure and function in the Zucker rat. J Lab Clin Med 106:598, 1985
53. O'Donnell MP, Kasiske BL, Cleary MP, Keane WF: Effects of genetic obesity on renal structure and function in the Zucker rat. II. Micropuncture studies. J Lab Clin Med 106:605, 1985
54. Witiak DT, Newman HAI, Fuller DR: Clofibrate and Related Analogs. Marcel Dekker, New York, 1977
55. Wasserman J, Santiago A, Holthofer H et al: Interactions of low density lipoprotein with rat mesangial cells. Kidney Int 35:1168, 1989
56. Gröne H-J, Walli AK, Gröne E et al: Receptor mediated uptake of apo B and apo E rich lipoproteins by human glomerular epithelial cells. Kidney Int 37:1449, 1990
57. Keane WF, Phillips J, Kasiske BL et al: Injurious effects of low density lipoprotein (LDL) on human mesangial cells (MC), abstracted. Kidney Int 37:509, 1990
58. Mulcahy WS, Smith JB, Selak M, Krishna GG: Cytotoxicity of low density lipoprotein (LDL) in cultured rat mesangial cells (MC), abstracted. Kidney Int 37:514, 1990
59. Wheeler DC, Persaud JW, Fernando R et al: Effects of low-density-lipoproteins on mesangial cell growth and viability in vitro. Nephrol Dial Transplant 5:181, 1990
60. Groggel GC, Cheung AK, Ellis-Benigni K, Wilson DE: Treatment of nephrotic hyperlipoproteinemia with gemfibrozil. Kidney Int 36:266, 1989
61. Utsunomiya K, Kurata H, Yoshizawa N et al: The preventive effect of HMG-CoA reductase inhibitor on diabetic nephropathy, abstracted. Diabetes suppl. 1 39:152A, 1990
62. O'Donnell MP, Kasiske BL, Schmitz PG, Keane WF: Effect of lovastatin (L) on progression of established glomerular disease, abstracted. J Am Soc Nephrol 1:638, 1990
63. Rabelink AJ, Hené RJ, Erkelens DW et al: Partial remission of nephrotic syndrome in patient on long-term simvastatin. Lancet 335:1045, 1990
64. Diamond JR, Hanchak NA, McCarter MD, Karnovsky MJ: Cholestyramine resin ameliorates chronic aminonucleoside nephrosis. Am J Clin Nutr 51:606, 1990

65. Schapel GJ, Edwards KD, Neale FC: Factorial study of the efficacy of cholestyramine, L-tryptophan and clofibrate in human nephrotic syndrome. Prog Biochem Pharmacol 9:82, 1974
66. Valeri A, Gelfand J, Blum C et al: Treatment of the hyperlipidemia of the nephrotic syndrome: a controlled trial. Am J Kidney Dis 8:388, 1986
67. Rabelink AJ, Hené RJ, Erkelens DW et al: Effects of simvastatin and cholestyramine on lipoprotein profile in hyperlipidaemia of nephrotic syndrome. Lancet 2:1335, 1988
68. Blantz RC, Wilson CB, Parthasarathy S, Bird JE: Antioxidant treatment and glomerular immune injury in the rat, abstracted. Kidney Int 37:477, 1990
69. Kaplan R, Aynedjian HS, Schlondorff D, Bank N: Renal vasoconstriction caused by short-term cholesterol feeding is corrected by thromboxane antagonist or probucol. J Clin Invest 86:1707, 1990
70. Iida H, Izumino K, Asaka M et al: Effect of probucol on hyperlipidemia in patients with nephrotic syndrome. Nephron 47:280, 1987
71. Schmitz PG, O'Donnell MP, Kasiske BL, Keane WF: Glomerular capillary pressure is increased in old obese Zucker rats and is unchanged with enalapril, abstracted. Kidney Int 37:520, 1990
72. Diamond JR, Karnovsky MJ: Exacerbation of chronic aminonucleoside nephrosis by dietary cholesterol supplementation. Kidney Int 2:671, 1987
73. Kelley VE, Izui S: Enriched lipid diet accelerates lupus nephritis in NZBxW mice: synergistic action of immune complexes and lipid in glomerular injury. Am J Pathol 111:288, 1983
74. Hurd ER, Mohnston JM, Okita JR: Prevention of glomerulonephritis and prolonged survival in New Zealand black/New Zealand white F_1 hybrid mice fed an essential fatty acid-deficient diet. J Clin Invest 67:476, 1981
75. Lefkowith JB, Schreiner G: Essential fatty acid deficiency depletes rat glomeruli of resident macrophages and inhibits angiotensin II-induced eicosanoid synthesis. J Clin Invest 80:947, 1987
76. Kelley VE, Ferretti A, Izui S, Strom T: A fish oil diet rich in eicosapentaenoic acid reduces cylooxygenase metabolites and suppresses lupus in MRL-1 mice. J Immunol 134:1914, 1984
77. Prickett JD, Robinson DR, Steinberg AD: Dietary enrichment with the polyunsaturated fatty acid eicosapentaenoic acid prevents proteinuria and prolongs survival in NZB $\times$ NZW F_1 mice. J Clin Invest 68:556, 1981
78. Robinson DR, Prickett JD, Makoul GT: Dietary fish oil reduces progression of established renal disease in (NZW/NZB) F_1 mice and delays renal disease in BXSB and MRL1 strains. Arthritis Rheum 29:539, 1986
79. Kher V, Barcelli U, Weiss M: Protective effect of polyunsaturated fatty acid supplementation in apoferritin induced murine glomerulonephritis. Prostaglandins Leukot Med 22:323, 1986
80. Ishimitsu T, Uehara T, Ishii M et al: Thromboxane and vascular smooth muscle cell growth in genetically hypertensive rats. Hypertension 12:46, 1988
81. Huttner JJ, Gwebu ET, Panganamala RV et al: Fatty acids and their prostaglandin derivatives: inhibitors of proliferation in aortic smooth muscle. Science 197:289, 1977
82. Menè P, Dunn MJ: Prostaglandins and rat mesangial cell proliferation. Kidney Int 37:1256, 1990
83. Endres S, Cahorbani R, Kelley VE et al: The effect of dietary supplementation with n-3 polyunsaturated fatty acids on the synthesis of interleukin-1 and tumor necrosis factor by mononuclear cells. N Engl J Med 320:265, 1989

84. Fox PL, DiCorletto PE: Fish oils inhibit endothelial cell production of platelet-derived growth factor-like protein. Science 241:453, 1988
85. Cartwright IJ, Pockley AG, Galloway JH et al: The effects of dietary ω-3 polyunsaturated fatty acids on erythrocyte membrane phospholipids, erythrocyte deformability and blood viscosity in healthy volunteers. Atherosclerosis 55:267, 1985
86. Donadio JV, Holman RT, Holub BJ, Bergstralh EJ: Effects of omega (ω)-3 polyunsaturated fatty acids (PUFA) in mesangial IgA nephropathy, abstracted. Kidney Int 37:255, 1990
87. Cheng IKP, Chan PCK, Chan MK: The effect of fish-oil dietary supplement on the progression of mesangial IgA glomerulonephritis. Nephrol Dial Transplant 5:241, 1990
88. Barcelli UO, Weiss M, Pollack VE: Effects of dietary prostaglandin precursor on the progression of experimentally induced chronic renal failure. J Lab Clin Med 100:786, 1982
89. Izumi Y, Weiner TW, Franco-saenz R, Mulrow PS: Effects of dietary linoleic acid on blood pressure and renal function in subtotally nephrectomized rats. Proc Soc Exp Biol Med 183:193, 1986
90. Heifets M, Morrissey JJ, Purkerson ML et al: Effect of dietary lipids on renal function in rats with subtotal nephrectomy. Kidney Int 32:335, 1987
91. Barcelli UO, Miyata J, Ito Y, Pollack V: Beneficial effect of polyunsaturated fatty acids in partially nephrectomized rats. Prostaglandins 32:211, 1986
92. Clark WF, Parbtani A, Philbrick D et al: Comparative efficacy of dietary treatments on renal function in rats with sub-total nephrectomy: renal polyunsaturated fatty acid incorporation and prostaglandin excretion. Clin Nephrol 33:25, 1990
93. Scharschmidt LA, Gibbons NB, McGarry L et al: Effects of dietary fish oil on renal insufficiency in rats with subtotal nephrectomy. Kidney Int 32:700, 1987
94. Anderson S, Meyer TW, Rennke HG, Brenner BM: Control of glomerular hypertension limits glomerular injury in rats with reduced renal mass. J Clin Invest 76:612, 1985
95. Nath KA, Kren SM, Hostetter TH: Dietary protein restriction in established renal injury in the rat. Selective role of glomerular capillary pressure in progressive glomerular dysfunction. J Clin Invest 78:1199, 1986
96. Schmitz PG, O'Donnell MP, Kasiske BL, Keane WF: Effects of dietary fatty acid supplementation on glomerular hemodynamics, abstracted. Kidney Int 37:556, 1990
97. Logan JL, Michael UF, Benson B: Effects of dietary fish oil on renal growth and function in uninephrectomized rats. Kidney Int 37:57, 1990
98. Adam O, Wolfram G: Effects of different linoleic acid intakes on prostaglandin biosynthesis and kidney function in man. Am J Clin Nutr 40:763, 1984
99. Bernardi P, Ghezzi F, Bastagli L et al: Effects of polyunsaturated fatty acids and prostaglandin synthesis on renal function. Prostaglandins Leukot Med 22:235, 1986
100. Düsing R, Struck A, Göbel BO et al: Effects of n-3 fatty acids on renal function and renal prostaglandin E metabolism. Kidney Int 38:315, 1990
101. Schaap GH, Bilo HJG, Popp-Snidjers C et al: Effects of protein intake variation and omega-3 polyunsaturated fatty acids on renal function in chronic renal disease. Life Sci 41:2759, 1987

8

The Role of Dietary Fatty Acids in Progressive Renal Disease

George F. Schreiner
Brad H. Rovin

INTRODUCTION

The potential contributing role of dietary fatty acids to the progression of renal disease and the complementary question of the protective effects of dietary lipid manipulation have emerged as significant areas of research only within the last several years. Dietary manipulation as a form of renal protective therapy has, for over half a century, emphasized restriction of protein, phosphate, or a caloric intake as antagonizing the evolution of nephrosclerosis in the damaged kidney.[1] Dietary fatty acid manipulation was initially studied in experimental immunologic models of glomerular injury. This was a result of accumulating evidence that products of fatty acid metabolism, via the cyclooxygenase or lipoxygenase pathway, could not only modulate the expression of humoral or cellular immunity, but also, when produced in the kidney, directly mediate much of the altered physiology of acute inflammation in the glomerulus.[2] The second significant set of observations that attracted the interest of experimental renal pathologists was the finding that hyperlipidemia exacerbates the progressive sclerosis of a variety of nonimmune renal diseases, including diabetic nephropathy and chronic, nephrotic-range proteinuria both in animal models and in human pathology.[3–5] The gap between these sets of observations has been bridged by the recent demonstrations that disease states lacking an immune etiology nonetheless have an immunologic component in the form of leukocytic infiltrates. This has suggested that lipid metabolism in the kidney, in a variety of disease states both immunologic and nonimmunological in origin, may serve as a mediating principle between the renal response to injury and its susceptibility to nephrosclerosis, which in many instances is mediated not only by alterations in renal filtration and blood flow but also by the intrarenal activation of cellular components of the immune system.

DIETARY FATTY ACIDS

The principal therapeutic manipulations of renal dietary fatty acids have been: (1) supplementation with fish oil, (2) supplementation with linoleic acid, the metabolic precursor of arachidonate, and (3) exclusion from the diet of both linoleic and arachidonic acids, the essential fatty acids.

The initial context with which these manipulations were undertaken was that of modulation of prostaglandin metabolism, which will be reviewed briefly. It should be borne in mind, however, that some of the effects that will be described in this review of fatty acid manipulation on the progression of renal disease have little or nothing to do with cyclooxygenase or lipoxygenase metabolism of arachidonate. The additional pathways implicated in the effects of these manipulations constitute a new direction in current investigations into lipid-dependent renal pathology.

Linoleate and arachidonate are essential fatty acids. For mammals, their requirements are absolute because mammals cannot desaturate fatty acids distal to the omega-9 carbon; both are omega-6 fatty acids. The numbering is based on the distance of the double bond from the methyl, or omega, ends of the fatty acids, which are additionally designated by the length of the carbon chain. Linoleate (18:2, omega-6) is an omega-6, 18-carbon fatty acid with two double bonds that undergoes elongation and additional desaturation to yield arachidonate (20:4, omega-6), which has two more carbons and four double bonds. Arachidonate is a component of the phospholipids and triglycerides of all tissues. When it is released from membranes by the action of phospholipase A_2, in reponse to a variety of signals, it is further oxidized via the cyclooxygenase and lipoxygenase pathways to yield prostaglandins and leukotrienes, respectively.[6] Fish oils consist, in part, of omega-3 fatty acids, with one more double bond than their omega-6 fatty acids of equivalent length. The fish oil counterpart of linoleic acid is linolenic acid, which can proceed through the same metabolic steps to become eicosapentaenoic acid (EPA); this acid is analogous to arachidonate, but has five double bonds instead of four. Eicosapentaenoic acid can be further oxidized via cyclooxygenase or lipoxygenase to generate metabolites whose biologic effects may be diminished from, or antagonistic to, the arachidonate-derived counterparts. Marine lipids thus compete with arachidonate for metabolic transformation.

In the dietary absence of linoleate or arachidonate, a biochemical event somewhat analogous to that of fish oil supplementation occurs. Oleate, an endogenously synthesized, 18-carbon fatty acid with one double bond (18 : 1, omega-9), is elongated and further desaturated to form eicosatrienoic acid (also known as Mead acid), which has the same carbon length as arachidonate, but one less double bond.[7] Mead acid is incorporated into cellular lipids in lieu of arachidonate. Essential fatty acid (EFA) deficiency need not be absolute. Its biochemical definition consists of a ratio of the phospholipid content of Mead acid to arachidonate exceeding 0.4.[8] EFA deficiency is a state of substrate deficiency, resulting in the naturally occurring products of cyclooxygenase and lipoxygenase becoming markedly diminished. Other pathways involving arachidonate turnover are similarly depressed, as will be reviewed below.

In contrast, linoleic acid supplementation augments the production of arachidonate metabolites. However, inasmuch as the products of a similar pathway may exert opposing effects, linoleate supplementation may exert entirely different effects depending on which metabolic pathway is preferentially activated. Prostaglandin E_2 (PGE_2), for example, increases renal blood flow while thromboxane A_2 (TXA_2) restricts it. Both are products of cyclooxygenase activity with differential kinetics of synthesis during the course of glomerulonephritis.[2]

EFA deficiency is induced by the removal of linoleate, arachidonate, or of all fat from the diet. Linoleate supplementation is commonly induced by the addition to the diet of the oil of plant seeds, such as sunflower or safflower oil. Fish oil supplementation occurs via the addition to the diet of oils extracted from cold-water fish, particularly that of salmon or menhaden, or by the administration of purified components of fish oil—linolenic acid, EPA, or docosahexanoic acid (DHA). Beef tallow is often used in studies as a source of saturated fat.

MODULATION OF RENAL INJURY BY DIETARY FISH OIL SUPPLEMENTATION

The use of dietary marine lipid supplementation as a therapeutic modality has been evaluated in a variety of progressive renal diseases. The most extensively studied process has been the immune complex glomerulonephritis (GN) of genetic models of murine lupus. The deposition of DNA–anti-DNA complexes in these animals produces a GN that ultimately leads to sclerosis and interstitial fibrosis. The two most commonly used models of spontaneous lupus nephritis are the MRL/lpr and NZB/NZW F_1 mice. Kelly et al. placed groups of lupus-prone MRL/lpr mice on standard laboratory chow or on diets containing 20 percent fish oil or safflower oil as the source of lipids.[9] The fish oil diet was high in EPA and DHA, both omega-3 fatty acids, while the safflower oil diet was enriched with linoleic acid. The animals maintained on the fish oil diet demonstrated a mean survival time 2 to 3 months greater than animals on lab chow or safflower diets. Significant proteinuria developed in 65 percent of the animals on the lab chow or safflower oil diets by 4 months of age compared with 7 percent of the mice consuming fish oil. Histologically, this correlated with less glomerular hypercellularity and IgG deposition and an absence of interstitial mononuclear cell infiltrates in the group receiving fish oil. The animals consuming fish oil had other indications of decreased immune activity, such as a reduction in lymphadenopathy, attenuated peritoneal macrophage Ia expression, and decreased levels of circulating gp70 immune complexes. It should be noted that the fish oil-enriched diet did not provide permanent protection from GN, but delayed its onset significantly. Linoleate supplementation had no effect on the evolution of the GN. Similar beneficial effects were observed by Robinson and colleages, who used three different strains of lupus-prone mice.[10–12] In their studies, mice were fed either a fish oil-supplemented diet or a control diet containing beef tallow as the lipid source. MRL/lpr mice placed on the fish oil diet tended to have a reduced mortality rate compared with the beef tallow group, although this did not reach statistical significance. The fish oil group did have significantly less proteinuria, glomerular basement membrane (GBM) thickening, and glomerular hypercellularity than the control group.[10] B/SB mice placed on the fish oil diet showed a marked increase in survival compared with the beef tallow group (no mortality at 13 months versus 90 percent mortality, respectively), a tendency toward less proteinuria, and significantly less glomerular sclerosis, GBM thickening, and glomerular hypercellularity.[10] Using the NZB/NZW F_1 lupus-prone mouse, the same investigators demonstrated that survival was prolonged with marine lipid supplementation whether the animals were started on dietary modification as weanlings[11] or at a mean age of 7.5 months, when proteinuria (and, thus, presumably renal disease) had already developed.[10] When placed on the diet as weanlings, all of the beef tallow group and none of the fish oil group developed proteinuria by 13.5 months of age; after renal disease had become manifest, 66 percent of the animals receiving beef tallow supplements progressed to heavy proteinuria compared with only 25 percent of the fish oil group.[10] Proteinuric animals subsequently started on fish oil had a sixfold decrease in the percent of scarred glomeruli at the time of death, but no differences in GBM thickening, hypercellularity, or interstitial infiltrates relative to the beef tallow group.[10] In an effort to establish a dose-response relationship between mortality and the amount of fish oil added to the diet of the NZB/NZW mice, Robinson's group found a marked increase in survival of animals on diets containing 25 or 11 percent fish oil, marginal benefits in animals receiving 5.5 percent fish oil, and no protection with 2.5 percent fish oil.[12]

Dietary fish oil supplementation has been used in other forms of GN. A 20 percent fish oil supplement was compared with a beef tallow diet alone or beef tallow plus safflower oil (to achieve high linoleic acid levels) in mice made nephritic by the repeated injection of apoferritin to induce correlating immune complexes.[13] The animals receiving either fish oil or the high linoleic acid diet demonstrated serum creatinine levels that were significantly less than the beef tallow group and that were not different from control, non-nephritic mice. Proteinuria was less in these groups, the protective effect of fish oil being somewhat greater than the high linoleic acid diet. As this model of GN is fairly mild, fish oil conferred no clearcut histologic benefit, although the mice fed fish oil tended to have less glomerular hypercellularity.

A fish oil diet was associated with marginal improvement compared with a beef tallow diet in the Adriamycin-induced nephrotic syndrome of rats.[14,15] The animals were started on the experimental diets after proteinuria became manifest. Rats receiving 14 percent fish oil tended to have less urinary protein (although still well above normal levels) and slightly lower levels of serum creatinine. No benefits in survival or glomerular histology were obtained; the fish oil group had less evidence of tubular injury than animals fed beef tallow.

Fish oil supplementation has been examined in the context of a non-immune model of progressive renal injury, subtotal renal ablation. Removal of one kidney and more than half of the single remnant kidney induces a syndrome of glomerular overwork resulting in proteinuria, glomerular hypertrophy, and eventual glomerulosclerosis (reviewed in Klahr et al.[16]). Two studies examining the effect of fish oil enrichment in partially nephrectomized rats have reached opposite conclusions. Scharschmidt et al. performed $\frac{4}{6}$ or $\frac{5}{6}$ nephrectomies on rats and maintained the animals on lab chow for 5 weeks before switching them to either a beef tallow or fish oil dietary formulation.[17] Nephrectomized ($\frac{5}{6}$) rats receiving fish oil had a more rapid rise in creatinine and a length of survival 2 months shorter than animals on the beef tallow diet. There was no difference in blood pressure, weight, or proteinuria between the groups. In rats undergoing a $\frac{4}{6}$ nephrectomy, proteinuria was worse in the fish oil group, but creatinine was similar to rats receiving beef tallow. Fish oil-fed animals manifested a slight increase in renal plasma flow, but a 25 percent decrease in glomerular filtration rate, resulting in a markedly decreased filtration fraction. Histologically, the glomeruli of the animals in the fish oil group displayed a more prominent degree of global sclerosis, although the glomeruli of both dietary groups showed evidence of focal sclerosis.

In contrast, two studies tested a variety of polyunsaturated fatty acid supplements in $\frac{3}{4}$-nephrectomized rats and found marine lipid supplementation to be beneficial.[18,19] There was no survival difference between beef tallow or fish oil groups, but the fish oil-fed animals had decreased proteinuria and significantly less glomerular sclerosis than control animals.[18] Two differences from the previous study were the placement on the diets immediately after surgery (before the onset of injury in the remnant kidney) and the use of salmon oil as the source of EPA as opposed to menhaden oil,[17] resulting in dietary compositional differences with respect to arachidonate and linoleate. Other differences between the studies included the development of diarrhea in the salmon oil-fed rats, potentially resulting in malabsorption of other nutrients that may have a role in progressive renal disease, and a small reduction of urinary PGE_2 excretion in the salmon oil group compared with a significantly reduced renal production of this prostaglandin in the study of Scharschmidt et al. The study of Barcelli et al. included multiple antioxidants in the diet; Scharschmidt et al.'s study did not, suggesting that oxidized fish oils may be toxic or that their beneficial effects may be neutralized by oxidation.

A more recent study examined the effects of 6 percent protein restriction, fish oil diet, or safflower oil diet started 1 week prior to $\frac{5}{6}$ nephrectomy and followed for 28 days. All three diets prevented the γ-globulinuria normally seen on a control diet, but only fish oil lessened the albuminuria. Fish oil and, to a lesser extent, safflower oil exhibited an enhanced ratio of prostacyclin to thromboxane in the urine. No histologic assays were done; there was no effect on blood pressure or renal clearance induced by the diets.

Another recent study examined the acute response of uninephrectomized rats on a fish oil-supplemented diet and observed increased glomerular filtration and renal blood flow with an inhibition of compensatory renal hypertrophy.[20] Current theories concerning the toxic effects of renal adaptation to injury indicate that this combination would increase susceptibility to glomerular sclerosis.[16] In an acute model of perfusion injury to the medullary thick ascending limb, fish oil exacerbated the severity and extent of tubular injury via its suppression of PGE_2 and PGI_2 and consequent exacerbation of renal ischemia.[21,22]

Cyclosporine (CSA) nephrotoxicity provides an additional model of hemodynamically mediated renal injury. Acute renal dysfunction during CSA administration results in part from afferent arteriolar vasoconstriction, possibly secondary to CSA-enhanced TXA_2 production, coupled with a decrease in prostacyclin activity. Rats receiving CSA orally developed a marked decline in glomerular filtration rate as measured by inulin clearance, proximal tubular vacuolization, and increased renal cortical TXB_2 synthesis. Administration of CSA in a fish oil vehicle instead of the usual olive oil vehicle attenuated the decrease in inulin clearance, the proximal tubular injury, and decreased TXB_2 synthesis below control levels in the kidney.[23] Rats treated with CSA were observed to have acute elevations in serum triglycerides and cholesterol. The concomitant administration of fish oil attenuated the rise in triglycerides by 50 percent and prevented an increase in cholesterol.[24] In both of these studies, fish oil-supplemented animals tended to have lower CSA levels than animals receiving CSA in olive oil; however, the CSA was always well above the therapeutic range.

There are few studies on the effect of dietary fish oil supplementation on the course of renal disease in humans. A small group of patients with IgA nephropathy received 1.6 g of EPA and 1 g of DHA daily for 1 year and were compared with a group of patients maintained on their normal diets.[25] Despite both groups having similar clinical parameters at the beginning of the study, renal function, as assessed by the reciprocal of the serum creatinine, deteriorated in the control group but remained stable in the fish oil group. Other investigators altered the diets of 15 patients with chronic renal insufficiency from a variety of causes, and assessed renal function at the end of each dietary period.[26] Diets consisting of low protein, high protein, and high protein plus 0.5 g of EPA daily were compared. The glomerular filtration rate and renal plasma flow tended to be highest when the patients were receiving EPA. However, there was no description of renal functional parameters prior to dietary manipulation for reference. Finally, the effect of fish oil supplementation was studied in 24 renal transplant recipients receiving CSA.[27] Patients were randomized to 6 g of fish oil or corn oil (high in linoleic acid) supplements daily for 3 months. Renal functional parameters were tested and compared with baseline values prior to dietary modification. Compared with baseline, the fish oil-supplemented group demonstrated a 20 percent increase in glomerular filtration rate, a 16.4 percent increase in renal plasma flow, a 21 percent fall in renal vascular resistance, and a decrease in mean arterial pressures and serum creatinine by 8.6 and 7.1 percent, respectively. The corn oil-fed controls did not have a significant change from baseline.

MODULATION OF RENAL INJURY BY DIETARY LINOLEATE SUPPLEMENTATION

The effect of dietary linoleate has been evaluated with respect to the progression of two experimental models of immune glomerular disease. In rats, linoleate supplementation of the diet significantly decreased but did not abolish the proteinuria and crescent formation induced in a model of accelerated anti-GBM nephritis, in which rats are preimmunized to heterologous antisera directed against the rat basement membrane.[28] Unfortunately, the control group was placed on a high linolenic diet derived from fish oil. There was no control group of rats on a standard laboratory diet with which to compare the experimental groups. The fish oil-treated animals displayed greater uremia and more crescent formation; there were no explanations for the exacerbation of the course of this lesion in fish oil-treated animals. There was no assessment as to whether the diets modulated the production of antibody or affected the inflammatory process per se at the level of the glomerulus. In a model of immune complex GN induced in mice via repeated innoculations with apoferritin, a high linoleic acid diet protected against the development of uremia and diminished, but did not abolish, proteinuria and the severity of histologic damage.[29] Again, the differential effects of the diets on the expression of humoral immunity or the glomerular response to the deposition of immune complexes was not delineated in these studies.

Dietary supplementation with linoleate has proven most effective in the model of glomerular sclerosis secondary to renal ablation. Barcelli and colleagues[30] first reported that a safflower oil diet protected against glomerulosclerosis and blunted the increases in blood urea nitrogen and proteinuria observed in rats with subtotal nephrectomies. Two other groups[31,32] have confirmed the histologic protection conferred by the diet, further observing augmentations in glomerular filtration rate and lower systemic blood pressure. Clark and colleagues observed that a safflower oil diet decreased γ-globulinuria and enhanced urinary content of PGI_2 metabolites in $\frac{5}{6}$ nephrectomized animals on the diet for 28 days, without much effect on blood pressure or renal function.[19] The remnant kidneys of rats fed a high linoleic acid diet demonstrate an increase in the renal content of linoleate.[32] The glomeruli from such animals express a marked increase in the spontaneous PGE_2 synthesis and release, particularly in conjunction with a low protein diet.[33] An acute model of renal disease demonstrating protection by dietary linoleate is the nonimmune injury to the medullary thick ascending limb induced by ischemia.[22]

MODULATION OF RENAL INJURY BY DIETARY RESTRICTION OF ESSENTIAL FATTY ACIDS

The unusual, renal-protective properties of a diet deficient in EFAs were first noted by Hurd et al.,[34] who were examining the anti-inflammatory effects of PGE_1 on the progression of renal insufficiency in the NZB/NZW F_1 strain of mouse, a murine model of systemic lupus erythematosus. To examine whether endogenous prostaglandin production may provide endogenous inhibition of the progression of GN, they placed mice on an EFA-deficient diet, reasoning that the diet should accelerate the development of renal insufficiency by inhibiting PGE production. Instead, the EFA-deficient state conferred remarkable protection against end-stage renal disease. The animals did not die of the GN and interstitial fibrosis that their littermates experienced on a controlled diet. Yet, the animals were not immunosuppressed, in the traditional sense of the term. The EFA-deficient mice had circulating immune complexes of DNA–anti-DNA, and deposited immune complexes and complement in the glomeruli. Other laboratories have demonstrated normal, delayed-type hypersensitivity reactions in EFA deficiency, including skin graft rejection and autoimmune reactivity to autologous antigen in adjuvant.[35,36] To date, there is little evidence that EFA deficiency significantly affects B- or T-cell reactivity to antigens. Our laboratory, in collaboration with James Lefkowith, considered a third possibility: that lipid restriction may affect the biology or reactivity of the renal macrophage. This possibility was suggested by the established effects of inhibited wound healing and granuloma formation observed in EFA-deficient rodents.[37,38] Both reactions are dependent on the recruitment of monocytes to the affected site and their subsequent activated transformation into macrophages.

We observed that EFA deficiency induced a marked, spontaneous depletion of the glomerular and interstitial macrophages that are normal basal constituents of the mesangial region and the renal cortical connective tissue.[39] Selective administration of linoleic acid to EFA-deficient rats caused a spontaneous reseeding of both locations with monocytes; administration of linolenic acid could not induce monocyte repopulation of the kidney. Our laboratory had previously shown that the resident renal macrophages are highly allogeneic by virtue of their membrane expression of Ia molecules, encoded by the major histocompatibility locus.[40] Renal macrophages are also highly efficient in presenting antigen to lymphocytes, suggestive of a potential role for the initiation of immune responses within the kidney itself. The EFA-deficient kidneys, nearly completely depleted of macrophages, were proven to be remarkably nonimmunogenic when transplanted across a major histocompatibility barrier into recipients on a controlled diet.[41] As the transplanted kidneys normalized their lipid content, we observed a spontaneous repopulation of the glomeruli and interstitium with macrophages from the host, confirming a metabolic linkage between the lipid status of the kidney and its susceptibility to monocyte traffic.

We next examined the potential effect of fatty acid restriction on monocyte migration into the kidney in acute inflammation. We had previously demonstrated that infiltrating monocytes are a major component of the hypercellular glomerulus in a model of rat GN induced by the administration of heterologous antisera directed against the rat GBM.[42] This lesion has two phases. In the first, several hours after the injection of antiserum, the glomerulus is infiltrated by neutrophils. Within 12 to 24 hours, they are replaced by infiltrating monocytes that chronically occupy the glomerular mesangium. The lesion progresses to a chronic mesangioproliferative GN. The neutrophil infiltrate is complement dependent; the monocyte invasion of the glomerulus is not. Prior decomplementation of animals with cobra venom has no effect on the stimulus for monocyte migration into the mesangium, which has remained unknown.[42] When EFA-deficient rats receive anti-GBM antisera, they actually experience an augmented, transient neutrophil infiltration of their glomeruli. However, the monocyte infiltration of the glomerulus, characteristic of the chronic phase, is completely abrogated, resulting in markedly diminished proteinuria, restoration of renal blood flow and glomerular filtration, and normalization of the sodium retention, polyuria, and mild uremia that are the characteristic expressions of this form of GN in rats on normal diets.[43] Because of both the depleted substrate and inhibition of monocyte infiltration, the kidney demonstrates virtually complete inhibition of the augmented TXA_2 and leukotriene B_4 (LTB_4) productions that is normally seen in the first 1 to 2 days of this lesion.

The effect of EFA deficiency is not confined to the glomerulus. A model of nephrosis, induced in rats by the injection of the aminonucleoside of puromycin, is characterized by a significant, acute interstitial infiltrate of macrophages and lymphocytes.[44] EFA deficiency completely inhibits the macrophage infiltrate, which peaks 7 to 10 days after the administration of the aminonucleoside of puromycin and 1 to 3 days after the onset of nephrotic-range proteinuria.[45] Interestingly, EFA deficiency also reverses the severe decrement in renal blood flow observed in the acute phase of proteinuria, an effect directly attributable to its prevention of the interstitial infiltrate. Inhibiting the interstitial nephritis by radiation-induced leukopenia reproduces the sparing effect of EFA deficiency on renal blood flow. The mechanism of the monocyte-mediated vasoconstriction in acute nephrosis remains unknown; it is not responsive to inhibitors of vasoconstrictive prostaglandins or of angiotensin conversion.[45] After 2 weeks of nephrosis, the glomerulus experiences an infiltration of monocytes that reside in the expanded mesangium in a chronically activated state.[42] Diamond and coworkers have shown that this phenomenon results in the development of focal, segmental glomerular sclerosis months after the development of proteinuria. Importantly, placing rats on a fatty acid-deficient diet, during the first month of proteinuria only, markedly inhibited the development of focal sclerosis 3 months later.[46] The experimental and control groups of animals were comparable with respect to proteinuria, hyperlipidemia, and hypertension. The EFA-deficient animals, however, were completely protected from the several-fold increase in glomerular macrophages. For the first time, these observations placed lipid-dependent glomerular macrophage accumulation in the pathway for the development of focal sclerosis.

The fatty acid modulation of monocyte traffic is also observed in organs other than the kidney. Wound healing in the skin and gut is impaired in the EFA-deficient state.[37] Wound healing is a macrophage-dependent phenomenon, although the presence of macrophage in the granulation tissue was not specifically measured in those early studies. A mouse model of autoimmune diabetes induced by the administration of streptozotocin, in which insulitis is preceded by a monocyte infiltration into the pancreatic islets, is completely prevented by EFA deficiency.[47] Similarly, EFA deficiency confers marked protection against the spontaneous development of autoimmune diabetes and thyroiditis in the BB/Wor rat.[48]

MECHANISMS OF PROGRESSIVE RENAL DYSFUNCTION

Manipulation of dietary fatty acids may effect the generation of toxic stimuli at a systemic level, such as the production of immune complexes or the sustenance of hypertension, or may modulate the regional renal response to injury, ranging from glomerular production of eicosanoids to the release of chemotactic factors. Intraglomerular pathways of adaptation to injury have received increasing emphasis because of evidence that the kidney has a limited number of responses to injury, immune or otherwise, many of which are a response to the loss of nephron mass rather than a unique response tailored to the nature of the toxic event that resulted in the functional or structural removal of nephron number. The process of sclerosis in which glomerular and interstitial structures are effaced by extracellular matrix[49] is subserved by a variety of metabolic pathways. Some receiving recent emphasis[16] include glomerular hypertrophy, systemic and intraglomerular hypertension, and in situ production of growth factors such as growth hormone or epidermal growth factors.[50,51] Intraglomerular activation of platelets to yield platelet-derived growth factor and systemically active peptide hormones such as angiotensin II have also been implicated.[16] The mesangium, consisting of contractile cells resembling vascular smooth muscle and resident or infiltrating leukocytes, is emerging as the initiating locus for glomerular sclerosis. Increased passage of macromolecules into the mesangium, mesangial cell proliferation and synthesis of extracellular matrix, and leukocyte infiltration are the earliest harbingers of eventual sclerosis.[16] Agents or pathways that particularly effect the biology of the mesangium appear to predominate in the current understanding of the events underlying the progression of renal insufficiency.

One such pathway that must be detailed in the context of fatty acid modulation of progressive renal injury is that of eicosanoid metabolism. Numerous aspects of renal physiology are affected by arachidonate metabolites. Glomerular production of TXA_2, a potent vasoconstrictor,[2] is increased in both immune and nonimmune models of renal injury, including nephrotoxic serum nephritis,[52,53] lupus nephritis,[8] subtotal renal ablation,[54] Adriamycin-induced nephrosis,[55] and the normotensive nephropathy of the Dahl salt-sensitive rat.[56] Whether the augmented thromboxane production is produced by intrinsic renal cells or by infiltrating leukocytes has not been determined in these models. Inhibition of thromboxane synthesis via the use of selective inhibitors has been reported to decrease proteinuria in the sclerosis of the remnant kidney,[54] in Adriamycin nephrosis,[55] and in salt-sensitive nephropathy.[56] Inhibition of thromboxane synthesis protects against the decline in glomerular filtration in acute GN[52,53] and in progressive sclerosis.[54] In addition to its vascular effects, TXA_2 is a potent inducer of platelet aggregation, with the consequent release of vasoactive and growth-promoting factors.[2]

The vasodilatory prostaglandins (such as PGE_2) and prostacyclin (PGI_2) serve as homeostatic mechanisms to oppose the effects of TXA_2. They maintain blood flow and glomerular filtration in the setting of decreased renal profusion or chronic kidney disease. In addition to renal blood flow,[57] PGE and PGI_2 have been shown to regulate salt and water transport,[58] the renal release of adrenergic transmitters,[59] and renin secretion.[60]

With this as background, one can examine the potential mechanisms whereby dietary fatty acid alterations might ameliorate some of the factors contributing to progressive renal disease.

MOLECULAR MECHANISMS OF FISH OIL ADMINISTRATION

Marine lipids affect both immune and nonimmune factors that may contribute to the progression of renal injury. It should be remembered that none of the effects on these factors have been unequivocally shown to be principally responsible for the protection against renal injury previously summarized.

Fish oil metabolites, particularly EPA, replace arachidonate in the cellular membrane phospholipids, resulting in the synthesis of the trienoic and pentanoic counterparts of the dienoic prostaglandins and LTB_4.[61] A decreased ratio of arachidonate to EPA is associated with the diminished release of interleukin-1 and tumor necrosis factor from the peripheral blood cells of normal volunteers.[62] Interleukin-1 and tumor necrosis factor are leukocyte-derived cytokines with numerous effects on nonimmune cells, including renal cells, that could contribute to renal dysfunction. Interleukin-1 stimulates matrix production by glomerular epithelial cells[63] and is a potent cofactor, with platelet-derived growth factor, in stimulating mesangial and endothelial cell proliferation.[64,65] Interleukin-1 can regulate salt and water transport by both proximal and distal tubules.[66] It is produced in glomeruli in both rat and human GN.[67,68] Fish oil inhibits release of endothelial-derived platelet-derived growth factor and could thus inhibit mesangial cell proliferation via suppression of growth-promoting cytokines derived from endogenous renal cells.[69]

Dietary fish oil also inhibits LTB_4 synthesis by leukocytes, and thus inhibits neutrophil chemotaxis.[70] LTB_4 and other products of the lipoxygenase pathway have been shown to modulate glomerular filtration, arteriolar resistance, and mesangial contractility.[71] An additional potential lipid mediator of renal injury is platelet-activating factor (PAF), which can induce mesangial cell contraction, induce the synthesis of vasoactive eicosanoids by glomerular cells, and activate circulating platelets and infiltrating leukocytes.[72,73] Renal synthesis of PAF is inhibited by a fish oil diet.[74] The incorporation of EPA in platelet membranes with a decrease in arachidonate and consequent inhibition of the platelet aggregant, TXA_2, results in diminished collagen-induced platelet activation.[75,76]

Fish oil may exert its most potent effects by its interruption of TXA_2 synthesis via depletion of its arachidonate precursor and production of the biologically inactive analog, TXA_3, as has been observed in a murine model of lupus GN.[9] As noted above, inhibition of thromboxane production has been efficacious in preserving glomerular filtration and decreasing proteinuria while minimizing histologic damage in a variety of acute and chronic models of renal injury. The presence of EPA also appears to shift the ratio of spontaneous production of TXA_2 and PGE_2, resulting in asymmetric suppression of thromboxane synthesis, out of proportion to the depletion of arachidonate.[19,75,77,78] In human subjects ingesting fish oil, the decrease in TXB_2 and LTB_4 is associated with the production of less active trienoic prostaglandins, such as TXA_3 and PGI_3, and augmented LTB_5.[79,80] Thus, fish oil supplementation favors a state of enhanced renal vasodilatation.[76]

Among the systemic effects of omega-3 polyunsaturated fatty acid supplementation is the lowering of blood pressure,[81] which has been shown to be renal protective in the remnant model of renal sclerosis.[16] In humans placed on a fish oil diet, erythrocyte membranes display a higher degree of unsaturated fatty acids, resulting in increased membrane deformability and consequent decreases in blood viscosity and increases in blood flow.[80,82] The degree of lipid unsaturation and the position of double bounds are important factors that regulate cell membrane fluidity.[83] In other systems, increasing membrane fluidity by incorporating unsaturated fatty acids into cell membranes has been associated with alterations in receptor number and affinity, changes in permeability, and changes in membrane fusion properties, such as phagocytosis.[84] Since the substitution of EPA and DHA into membrane phospholipids results in displacement of linoleic acid,[83] fish oil supplementation theoretically increases the double-bond index of cell membranes and, thus, their fluidity. Additionally, long chain, polyunsaturated fatty acids have recently been shown to influence certain aspects of cellular immune function, such as cytotoxic T-cell lysis of targets[85,86] and neutrophil production of oxygen radicals.[87] It is thus conceivable that fish oil may effect cell activities directly at the membrane level as a function of the structural properties of its constituent fatty acids. The implications of this for renal adaptation to injury are unknown.

Given its anti-inflammatory effects, the direct inhibitions of immune parameters by fish oil are surprisingly few, particularly when examined under in vivo conditions. Interleukin-1 and tumor necrosis factor, whose release is inhibited by fish oil, are participatory factors in the clonal activation and differentiation of lymphocytes in response to antigen.[88] Fish oil diminishes spleen and lymph node weights in animals with lupus GN[9,12] and decreases macrophage expression of Ia,[9] indicating some degree of inhibition of immune activation. Yet, fish oil-fed animals do not exhibit global immunosuppression, nor is there any data to suggest such a state in humans. Thus, animals with lupus nephritis do not exhibit decreased levels of anti-DNA antibodies,[9,11] suggesting that the locus of efficacy of fish oil resides at the level of the glomerulus and the interstitium and not at the level of the lymph node.

Finally, some of the beneficial effects of fish oil may lie in its well-established capacity to lower serum triglycerides and cholesterol,[89] two recently recognized cofactors in exacerbating the development of nephrosclerosis in experimental diabetic nephropathy, in the remnant kidney and in a variety of human examples of renal insufficiency.[4,5] Both normal and hyperlipidemic humans as well as hyperlipidemic rats experience lowered serum cholesterol while on diets supplemented with fish oil.[90–93] Postulated mechanisms have included decreases in very low-density and low-density lipoprotein synthesis and enhanced low-density lipoprotein receptor expression.[93–95]

MOLECULAR MECHANISMS OF LINOLEATE SUPPLEMENTATION

As the precursor to arachidonate, linoleate might be expected to globally increase all arachidonate metabolites. In fact, for reasons not clearly understood, prostaglandins of the E series are preferentially synthesized in both humans and experimental animals placed on a high linoleate diet.[96,97] Renal production of PGE_2 is diminished in the remnant kidney, but augmented when the rats are placed on linoleate supplementation,[98] an effect that is enhanced by the additional restriction of protein in the diet.[33] There is little good evidence that linoleate supplementation exerts a significant immunosuppressive effect. In the study demonstrating a protective effect of linoleate on an immunization-induced immune complex GN, diminished IgG deposits were noted in the mouse glomeruli, but there was no evaluation of basic parameters of immune activation.[29] The progression of end-stage renal insufficiency in the NZB/NZW F_1 mouse is not affected by dietary linoleate supplementation,[9] suggesting a lack of effect on systemic immune activation. The well-known inhibitory effects of prostaglandins of the E series, particularly PGE_1, on in vitro expressions of T cell, B cell, and macrophage activation require pharmacologic concentrations of prostaglandins not encountered in vivo during the course of dietary supplementation.[99–102]

The higher systemic level of PGE production in linoleate-supplemented animals does appear to modulate platelet activity. Animals on this diet exhibit a prolonged bleeding time and antagonism of platelet aggregation induced by thrombin.[103] The enhanced TXA_2 production of immune GN[52] would be expected to exert a platelet-activating effect. Inhibition of thromboxane synthetase diminishes proteinuria and crescent formation in experimental GN, in part by reversing platelet hyperaggregatability.[104] Linoleate supplementation increases creatinine clearance and lowers blood pressure in humans[105,106] and lowers blood pressure and promotes sodium excretion in rats, consistent with augmentation of endogenous renal release of PGE.[107,108] Prostaglandin E_2, as it dilates renal blood flow, promotes perfusion of the inner cortex and medulla,[109,110] a region of the kidney most susceptible to ischemic injury.[22] PGE_2 also directly inhibits oxygen consumption by the medulla,[111] effectively blocking sodium chloride transport,[112] in part by inhibiting Na^+-K^+-ATPase.[113] Thus, the principal effects of the addition of linoleic acid to the diet appear to be mediated by the production of vasodilatory prostaglandins, the inhibition of platelet activation, and the inhibition of renal demand for oxygen under conditions of reduced renal blood flow. Whether there are additional cellular processes affected, independent of substrate-driven enhancement of cyclooxygenase activity, remains unknown.

MOLECULAR MECHANISMS OF ESSENTIAL FATTY ACID RESTRICTION

Essential fatty acid deficiency would predictably decrease the synthesis of both vasodilatory and vasoconstrictive prostaglandins by simple virtue of substrate depletion. While Mead acid can be metabolized via cyclooxygenase, the products are not known to exhibit significant biologic activity.[7] The synthesis of LTB_4 is additionally suppressed due to the formation of LTA_3 from Mead acid. LTA_3 is a suicide substrate, inactivating LTA hydrolase and preventing the metabolism of arachidonate to LTB_4.[114] In both the normal state and in acute GN, EFA deficiency markedly blunts the glomerular synthesis of PGE_2, TXA_2, and LTB_4.[39,43] The physiologic implications of this effect depends on the selective effects of whatever eicosanoid is contributing to a given lesion at a given point in time. The nature of renal arachidonate metabolism is to eventually generate a counterbalancing effect by the synthesis of an opposing metabolite in any given disease state expressing itself via the local generation of arachidonate products.[2] This effect, of course, would also be predictably blunted in EFA deficiency. Thus, it is difficult to assess a specific protective effect to EFA deficiency based solely on global interruption of the arachidonate pathway. The issue is not resolvable in the absence of the technical capability of reintroducing specific arachidonate metabolites in the EFA-deficient state.

Our studies have indicated that EFA deficiency affects another pathway that is independent of cyclooxygenase or lipoxygenase activity: a lipid-dependent mechanism for the initiation of a monocyte influx in the context of acute inflammation. Investigation of this pathway was prompted by the observation that EFA-deficient kidneys exhibit basal depletion of resident macrophages and ablated monocyte influx in GN, with neutrophil infiltration in this model of inflammation remaining unimpaired.[39,43] In experimental nephrosis, monocyte infiltration of the glomerulus and the interstitium is also specifically prevented.[45,46] There are two explanations for this effect: either EFA-deficient macrophages are impaired with respect to their capacity to respond to signals eminating from stressed tissue, or EFA-deficient tissue is inhibited with respect to its synthesis or release of a monocyte chemoattractant.

EFA-deficient macrophages are capable of chemotactic migration toward complement and toward platelet-activating factor.[43,115] There is no circulating inhibitor of monocyte chemotaxis in EFA-deficient serum.[43] Monocytes do not chemotactically migrate toward cyclooxygenase products, LTB_4, or 5-, 12-, or 15-hydroxyeicosatetraenoic acid.[115] Indeed, inhibition via indomethacin of cyclooxygenase activity actually augments monocyte infiltration of the glomerulus in both acute[115] and chronic[116] models of GN.

We have found that acutely inflamed glomeruli release a lipid chemoattractant activity that is specific for monocytes and has no effect on neutrophil migration. Chloroform–methanol extraction of the supernatants of cultured, nephritic glomeruli induces a partitioning of the activity into the organic phase. EFA-deficient, nephritic glomeruli yield considerably less activity, not significantly different from that obtained from control, uninflamed glomeruli.[115] The monocyte chemoattractant appears to be a complex, neutral lipid based on its chromatographic properties by high-performance and thin-layer chromatography. It appears to have free hydroxyl groups, but no free carboxy groups.[117] A similar, proinflammatory material is released by pancreatic islets oxidatively stressed by streptozotocin.[118] Its chromatographic properties and the lack of its inhibitory effect, in vivo, by administration of a receptor antagonist[115] indicate that the material is not platelet-activating factor.

We believe this to be a novel lipid pathway by which stressed or damaged tissue induces monocyte infiltration. EFA deficiency appears to exert its anti-inflammatory effect in the glomerulus and in the renal interstitium, and perhaps in the skin, pancreas, and thyroid, via suppression of the synthesis or release of this factor or factors. Clarification of the relationship between depletion of arachidonate or linoleic acid and the inhibited release of this chemotactic lipid awaits identification of the material. Once monocytes have invaded the kidney and are activated in situ, they are admirably equipped to contribute to sclerosis via the release of cytokines, such as interleukin-1 reactive oxygen species, growth factors (including platelet-derived growth factor), and proteolytic enzymes. All may contribute to promote altered glomerular and tubular physiology, infiltration and proliferation of fibroblast and vascular smooth muscle cell, excess or inappropriate synthesis of collagen and other components of the extracellular matrix, disruption of the integrity of basement membranes, and eventual obliteration of differentiated structures within the microenvironment of their influx.[119,120]

The mechanisms underlying the sustained presence of leukocytes within the renal glomerulus and interstitium and the regulation of their activation in situ comprise emerging salients in the development of a new understanding of the lipid contributions to the pathophysiology of progressive renal disease. The implications of the lipid mechanisms underlying exacerbation or slowing of nephrosclerosis will extend to classic immune renal disorders as well as to renal pathology characterized as having an immune cell component, including diabetes, obstructive uropathy, and chronic proteinuria.[120]

REFERENCES

1. Olson J, Heptinstall R: Nonimmunologic mechanisms of renal injury. p. 155. In Rubin E, Damjanov I (eds): Pathology Reviews. Humana Press, Clifton, NJ, 1989
2. Dunn M: Renal prostaglandins. p. 1. In Dunn M (ed): Renal Endocrinology. Williams & Wilkins, Baltimore, 1983
3. Schmitz PG, Kasiske BL, O'Donnell MP, Keane WF: Lipids and progressive renal injury. Semin Nephrol 9:354, 1989
4. Keane WF, Kasiske BL, O'Donnell MP: Lipids and progressive glomerulosclerosis: a model analogous to atherosclerosis. Am J Nephrol 8:261, 1988
5. Moorhead J, Wheeler D, Varghese Z: Permissive role of lipids in progressive renal disease. Am J Med 87:12N, 1989
6. Vance D, Vance J: Biochemistry of Lipids and Membranes. Benjamin/Cummings Publishing, Menlo Park, CA, 1985
7. Mead JF: The essential fatty acids: past, present, and future. Prog Lipid Res 20:1, 1982
8. Holman R: The ratio of trienoic:tetraenoic acids in tissue lipids as a measure of essential fatty acid requirement. J Nutr 70:405, 1960
9. Kelly VE, Feffetti A, Izui S, Strom TB: A fish oil diet rich in eicosapentaenoic acid reduces cyclooxygenase metabolites and suppresses lupus in MRL-lpr mice. J Immunol 134:1914, 1985
10. Robinson DR, Prickett JD, Makoul GT et al: Dietary fish oil reduces progression of established renal disease in (NZB×NZW)F_1 mice and delays renal disease in BXSB and MRL-lpr strains. Arthritis Rheum 29:539, 1986
11. Prickett JD, Robinson DR, Steinberg AD: Dietary enrichment with the polyunsaturated fatty acid eicosapentaenoic acid prevents proteinuria and prolongs survival in NZB×NZW F_1 mice. J Clin Invest 68:556, 1981
12. Robinson DR, Prickett JD, Polisson R et al: The protective effect of dietary fish oil on murine lupus. Prostaglandins 30:51, 1985
13. Kher V, Barcelli U, Weiss M et al: Protective effect of polyunsaturated fatty acid supplementation in apoferritin induced murine glomerulonephritis. Prostaglandins Leukot Med 22:323, 1986
14. Ito Y, Barcelli U, Yamashita W et al: Dietary fat in experimental nephrotic syndrome: beneficial effects of fish oil on serum lipids and indirectly on the kidney. Life Sci 40:2317, 1987
15. Ito Y, Barcelli U, Yamashita W et al: Fish oil has beneficial effects on lipids and renal disease of nephrotic rats. Metabolism 37:352, 1988
16. Klahr S, Schreiner G, Ichikawa J: The progression of renal disease. N Engl J Med 318:1657, 1988
17. Scharschmidt LA, Gibbons NB, McGarry L et al: Effects of dietary fish oil on renal insufficiency in rats with subtotal nephrectomy. Kidney Int 32:700, 1987

18. Barcelli UO, Miyata J, Ito Y et al: Beneficial effects of polyunsaturated fatty acids in partially nephrectomized rats. Prostaglandins 32:211, 1986
19. Clark W, Parbtani A, Philbrick D et al: Comparative efficacy of dietary treatments on renal function in rats with sub-total nephrectomy: renal polyunsaturated fatty acid incorporation and prostaglandin excretion. Clin Nephrol 33:25, 1990
20. Logan JL, Michael UF, Benson B: Effects of dietary fish oil on renal growth and function in uninephrectomized rats. Kidney Int 37:57, 1990
21. Kasiske B, O'Connell M, Keane W: Fish oil diet decreases prostacyclin and increases resistance in isolated rat kidneys. Am J Physiol 256:F843, 1989
22. Silva P, Rosen S, Spokes K et al: Influence of endogenous prostaglandins on mTAL injury. J Am Soc Nephrol 1:808, 1990
23. Elizinga L, Kelley VE, Houghton DC, Bennett WM: Modification of experimental nephrotoxicity with fish oil as the vehicle for cyclosporine. Transplantation 43:271, 1987
24. Jevnikar AM, Petric R, Holub BJ et al: Effect of cyclosporine on plasma lipids and modification with dietary fish oil. Transplantation 46:722, 1988
25. Hamazaki T, Tateno S, Shishido H: Eicosapentaenoic acid and IgA nephropathy. Lancet 1:1018, 1984
26. Schaap GH, Bilo HJG, Poep-Snijders C et al: Effects of protein intake variation and omega-3 polyunsaturated fatty acids on renal function in chronic renal disease. Life Sci 41:2795, 1987
27. Homan Van Der Heide JJ, Bilo HJG, Tegzezz AM, Donker AJM: The effects of dietary supplementation with fish oil on renal function in cyclosporine-treated renal transplant recipients. Transplantation 49:523, 1990
28. Watanabe S, Suzuki E, Kojima R et al: Effect of dietary α-linolenate/linoleate balance on crescent type anti-glomerular basement membrane nephritis in rats. Lipids 25:267, 1990
29. Kler V, Barcelli U, Weiss M, Pollak V: Effects of dietary linoleic acid enrichment on induction of immune complex nephritis in mice. Nephron 39:261, 1985
30. Barcelli UO, Weiss M, Pollak VE: Effects of dietary prostaglandin precursor on the progression of experimentally induced chronic renal failure. J Lab Clin Med 100:786, 1982
31. Izumi Y, Weiner TW, Franco-Saenz R, Mulrow PJ: Effects of dietary linoleic acid on blood pressure and renal function in subtotally nephrectomized rats. Proc Soc Exp Biol Med 183:193, 1986
32. Heifets M, Morrissey JJ, Purkerson ML et al: Effect of dietary lipids on renal function in rats with subtotal nephrectomy. Kidney Int 32:335, 1987
33. Ito Y, Barcelli U, Yamashita W et al: A low protein-high linoleate diet increases glomerular PGE_2 and protects renal function in rats with reduced renal mass. Prostaglandins Leukot Med 28:277, 1987
34. Hurd ER, Johnston JM, Okita JR et al: Prevention of glomerulonephritis and prolonged survival in New Zealand black/New Zealand white F_1 hybrid mice fed an essential fatty acid-deficient diet. J Clin Invest 67:476, 1981
35. Mertin J, Hunt R: Influence of polyunsaturated fatty acids on survival of skin allografts and tumor incidence in mice. Proc Natl Acac Sci USA 73:928, 1976
36. Selivonchick D, Johnston P: Fat deficiency in rats during development of the central nervous system and susceptibility to experimental, allergic encephalomyelitis. J Nutr 105:288, 1975

37. Hulsey T, O'Neill J, Neblett W, Meng H: Experimental wound healing in essential fatty acid deficiency. J Pediatr Surg 15:505, 1980
38. Bonta I, Parnham M, Adolfs M: Reduced exudation and increased tissue proliferation during chronic inflammation in rats deprived of endogenous prostaglandin precursors. Prostaglandins 14:295, 1977
39. Lefkowith J, Schreiner GF: Essential fatty acid deficiency depletes rat glomeruli of resident macrophages and inhibits angiotensin II-induced eicosanoid synthesis. J Clin Invest 80:946, 1987
40. Schreiner GF, Kiely J-M, Cotran RS, Unanue ER: Characterization of resident glomerular cells in the rat expressing Ia determinants and manifesting genetically restricted interactions with lymphocytes. J Clin Invest 68:920, 1981
41. Schreiner G, Flye W, Brunt E et al: Essential fatty acid depletion of renal allografts and prevention of rejection. Science 240:1032, 1988
42. Schreiner GF, Cotran RS, Unanue ER: Modulation of Ia and leukocyte common antigen expression in rat glomeruli during the course of glomerulonephritis and aminonucleoside nephrosis. Lab Invest 51:524, 1984
43. Schreiner GF, Rovin B, Lefkowith JB: The anti-inflammatory effects of essential fatty acid deficiency in experimental glomerulonephritis: the modulation of macrophage migration and eicosanoid metabolism. J Immunol 143:3192, 1989
44. Eddy AA, Michael AF: Acute tubulointerstitial nephritis associated with aminonucleoside nephrosis. Kidney Int 33:14, 1988
45. Harris K, Lefkowith J, Klahr S, Schreiner GF: Essential fatty acid deficiency ameliorates acute renal dysfunction in the rat following the administration of the aminonucleoside of puromycin. J Clin Invest 86:1115, 1990
46. Diamond J, Pesek I, Ruggieri S, Karnovsky M: Essential fatty acid deficiency during acute puromycin nephrosis ameliorates the late renal injury. Am J Physiol 257:F796, 1989
47. Wright JR, Lefkowith JB, Schreiner GF, Lacy PE: Essential fatty acid deficiency prevents multiple low-dose streptozotocin-induced diabetes in CD-1 mice. Proc Natl Acad Sci USA 85:6137, 1988
48. Lefkowith J, Schreiner G, Cormier J et al: Prevention of diabetes in the BB rat by essential fatty acid deficiency: relationship between physiological and biochemical changes. J Exp Med 171:729, 1989
49. Border WA, Okuda S, Nakamura T: Extracellular matrix and glomerular disease. Semin Nephrol 9:307, 1989
50. Fogo A, Ichikawa I: Evidence for the central role of glomerular growth promoters in the development of sclerosis. Semin Nephrol 9:329, 1989
51. Doi T, Striker L, Gibson C et al: Glomerular lesions in mice transgenic for growth hormone and insulinlike growth factor-1. Am J Pathol 137:541, 1990
52. Lianos EA, Andres GA, Dunn MJ: Glomerular prostaglandin and thromboxane synthesis in rat nephrotoxic serum nephritis. J Clin Invest 72:1439, 1983
53. Stork JE, Dunn MJ: Hemodynamic roles of thromboxane A_2 and prostaglandin E_2 in glomerulonephritis. J Pharmacol Exp Ther 233:672, 1985
54. Purkerson ML, Joist JH, Yates J et al: Inhibition of thromboxane synthesis ameliorates the progressive kidney disease of rats with subtotal renal ablation. Proc Natl Acad Sci USA 82:193, 1985
55. Remuzzi G, Imberti L, Rossini M et al: Increased glomerular thromboxane synthesis as a possible cause of proteinuria in experimental nephrosis. J Clin Invest 75:94, 1985

56. Yamashita W, Ito Y, Weiss MA et al: A thromboxane synthetase antagonist ameliorates progressive renal disease of Dahl-S rats. Kidney Int 33:77, 1988
57. Herbaczynska-Cedro K, Vane JR: Contribution of intrarenal generation of prostaglandins to autoregulation of renal blood flow in the dog. Circ Res 23:428, 1973
58. Grantham JJ, Orloff J: Effect of prostaglandin E_1 on the permeability response of the isolated collecting tubule to vasopressin, adenosine 3′-5′ monophosphate and theophylline. J Clin Invest 47:1154, 1968
59. Malik KV, McGiff JC: Modulation by prostaglandins of adrenergic transmission in the isolated perfused rabbit and rat kidney. Circ Res 36:599, 1975
60. Vander AJ: Direct effects of prostaglandin on renal function and renin release in anesthetized dogs. Am J Physiol 214:218, 1968
61. Barcelli U, Pollak V: Is there a role for polyunsaturated fatty acids in the prevention of renal disease and renal failure? Nephron 41:209, 1985
62. Endres S, Ghorbani R, Kelley VE et al: The effect of dietary supplementation with n-3 polyunsaturated fatty acids on the synthesis of interleukin-1 and tumor necrosis factor by mononuclear cells. N Engl J Med 320:265, 1989
63. Torbohm I, Berger B, Sconermark M et al: Modulation of collagen synthesis in human glomerular epithelial cells by interleukin 1. Clin Exp Immunol 75:427, 1989
64. Lovett DH, Ryan JL, Steizel RB: Stimulation of rat mesangial cell proliferation by macrophage interleukin 1. J Immunol 131:2830, 1983
65. Ooi BS, MacCarthy EF, Hsu A, Ooi YM: Human mononuclear cell modulation of endothelial cell proliferation. J Lab Clin Med 102:428, 1983
66. Schreiner GF, Kohan D: Regulation of renal transport processes and hemodynamics by macrophages and lymphocytes. Am J Physiol 258:F761, 1990
67. Matsumoko K, Dowling T, Atkins RC: Production of interleukin 1 in glomerular cell cultures from patients with rapidly progressive crescentic glomerulonephritis. Am J Nephrol 8:463, 1988
68. Matsumoto K, Hatano M: Production of interleukin 1 in glomerular cell cultures from rats with nephrotoxic serum nephritis. Clin Exp Immunol 75:123, 1989
69. Fox P, DiCorleto P: Fish oils inhibit endothelial cell production of platelet-derived growth factor-like protein. Science 241:453, 1988
70. Lee TH, Hoover RL, Williams JD et al: Effect of dietary enrichment with eicosapentaenoic and docasahexaenoic acids on in vitro neutrophil and monocyte leukotriene generation and neutrophil function. N Engl J Med 312:1217, 1985
71. Ardaillou R, Band L, Sraer J: Leukotriene and reactive oxygen species as mediators of glomerular injury. Am J Nephrol 9:17, 1989
72. Schlondorff D, Neuwirth R: Platelet-activating factor and the kidney. Am J Physiol 251:F1, 1986
73. Bertani T, Livio M, Macconi D et al: Platelet activating factor (PAF) as a mediator of injury in nephrotoxic nephritis. Kidney Int 31:1248, 1987
74. Young Y, Philbrick O, Holub B: The effect of long-term consumption of fish oil on platelet activating factor synthesis in rat renal microsomes. Biochem Biophys Res Commun 160:1238, 1989
75. Siess W, Roth P, Scherer B et al: Platelet membrane fatty acids, platelet aggregation, and thromboxane formation during a mackeral diet. Lancet 1:441, 1980

76. Needleman P, Raz A, Minkes MS et al: Triene prostaglandins: prostacyclin and thromboxane biosynthesis and unique biological properties. Proc Natl Acad Sci USA 76:944, 1979
77. Lorenz R, Spengler U, Fischer S et al: Platelet function, thromboxane formation and blood pressure control during supplementation of the Western diet with cod liver oil. Circulation 67:504, 1983
78. Von Schacky C, Fischer S, Weber PC: Long-term effects of dietary marine omega-3 fatty acids upon plasma and cellular lipids, platelet function, and eicosanoid formation in humans. J Clin Invest 76:1626, 1985
79. Von Schacky C: Prophylaxis of atherosclerosis with marine omega-3 fatty acids. Ann Intern Med 107:890, 1987
80. Clark WF, Parbtani A, Huff MW et al: Omega-3 fatty acid dietary supplementation in systemic lupus erythematosus. Kidney Int 36:653, 1989
81. Knapp HP, FitzGerald GA: The antihypertensive effects of fish oil: a controlled study of polyunsaturated fatty acid supplements in essential hypertension. N Engl J Med 320:1037, 1989
82. Cartwright IJ, Pockley AG, Galloway JH et al: The effects of dietary omega-3 polyunsaturated fatty acids on erythrocyte membrane phospholipids, erythrocyte deformability and blood viscosity in healthy volunteers. Atherosclerosis 55:267, 1985
83. Popp-Snijders C, Schonlen JA, DeJong AP, Van Der Veen SA: Effect of cod liver oil on the lipid composition of human erythrocyte membranes. Scand J Clin Lab Invest 44:39, 1984
84. Hagve TV: Effects of unsaturated fatty acids on cell membrane functions. Scand J Clin Lab Invest 48:381, 1988
85. Richieri GV, Kleinfeld AM: Free fatty acids inhibit cytotoxic T lymphocyte-mediated lysis of allogeneic target cells. J Immunol 145:1074, 1990
86. Richieri GV, Kleinfeld AM: Free fatty acid perturbation of transmembrane signaling in cytotoxic T lymphocytes. J Immunol 143:2302, 1989
87. Badwey JA, Curnutte JT, Robinson JM et al: Effects of free fatty acids on release of superoxide and on change of shape by human neutrophils. J Biol Chem 259:7870, 1984
88. Unanue ER, Allen PM: The basis for the immunoregulatory role of macrophages and of the accessory cells. Science 236:551, 1987
89. Mueller BA, Talbert RL: Biological mechanisms and cardiovascular effects of omega-3 fatty acids. Clin Pharmacol 7:795, 1988
90. Harris W, Connor W, McMurry M: The comparative reductions of the plasma lipids and lipoproteins by dietary polyunsaturated fats: salmon oil versus vegetable oils. Metabolism 32:179, 1983
91. Phillipson B, Rothrock D, Connor W: et al: Reduction of plasma lipids, lipoproteins, and apoproteins by dietary fish oils in patients with hypertriglyceridemia. N Engl J Med 312:1210, 1988
92. Barcelli U, Beach D, Thompson B et al: A diet containing n-3 and n-6 fatty acids favorably alters the renal phospholipid eicosanoid synthesis, and plasma lipids in nephrotic rats. Lipids 23:1059, 1988
93. Shepherd J, Packard CJ, Grundy SM et al: Effects of saturated and polyunsaturated fat diets on the chemical composition and metabolism of low density lipoproteins in man. J Lipid Res 21:91, 1980
94. Nestel P, Connor W, Reardon M et al: Suppression by diets rich in fish oil of very low density lipoprotein production in man. J Clin Invest 74:82, 1984

95. Illingworth D, Harris W, Connor W: Inhibition of low density lipoprotein synthesis by dietary omega-3 fatty acids in humans. Arteriosclerosis 4:270, 1984
96. Hwang DH, Mathias M, Dupont J, Meyer DL: Linoleate enrichment of diet and prostaglandin metabolism in rats. J Nutr 105:995, 1975
97. Epstein M, Lifschitz M, Rappaport K: Augmentation of prostaglandin production by linoleic acid in man. Clin Sci 63:565, 1982
98. Barcelli U, Weiss M, Pollak V: Effects of a dietary prostaglandin precursor on the progression of experimentally induced chronic renal failure. J Lab Clin Med 100:786, 1982
99. Pelus LM, Strausser HR: Prostaglandins and immune response. Life Sci 20:903, 1977
100. Goodwin JS, Bankhurst AD, Meesner RP: Suppression of human T cell mitogenesis by prostaglandin. Existence of a prostaglandin producing suppressor cell. J Exp Med 146:1719, 1977
101. Zimecki M, Webb DR: Regulation of immune response to T independent antigens and B cells. J Immunol 117:2158, 1976
102. Hwang D: Essential fatty acids and immune response. FASEB J 3:2052, 1989
103. McGregor L, Morazain R, Renaud S: Effect of dietary linoleic acid on platelet function in the rat. Thromb Res 20:499, 1980
104. Nagamatsu T, Tsukushi Y, Ito M et al: Antinephritic effect of *OKY-046,* a thromboxane A synthetase inhibitor. Jpn J Pharmacol 49:501, 1989
105. Adam O, Wolfram G: Effect of different linoleic acid intakes on prostaglandin biosynthesis and kidney function in man. Am J Clin Nutr 40:763, 1984
106. Comberg H, Heyden S, Hames C: Hypotensive effects of dietary prostaglandin precursors in hypertensive man. Prostaglandins 15:193, 1978
107. Grantham JJ, Orloff J: Effect of prostaglandin E_1 on the permeability response of the isolated collecting tubule to vasopressin, adenosine 3′-5′ monophosphate and theophylline. J Clin Invest 47:1154, 1968
108. TenHoor F, Van de Graaf H: The influence of a linoleic acid-rich diet and of acetyl salicylic acid on Nath-induced hypertension, Na^+ and H_2O-balance and urinary prostaglandin excretion in rats. Acta Biol Med Ger 37:875, 1978
109. Jackson E, Heidemann H, Branch R, Gerkens J: Low dose intrarenal infusions of PGE_2, PGI_2 and 6-keto-PGE_1 vasodilate the in vivo rat kidney. Circ Res 51:67, 1982
110. Kirschenbaum M, White N, Stein J, Ferris T: Redistribution of renal cortical blood flow during inhibition of prostaglandin synthesis. Am J Physiol 227:801, 1974
111. Lear S, Silva P, Kelley V, Epstein F: Prostaglandin E_2 inhibits oxygen consumption in rabbit medullary thick ascending limb. Am J Physiol 258:F1372, 1990
112. Stroke J: Effect of prostaglandin E_2 on chloride transport across the rabbit thick ascending limb of Henle. Selective inhibition of the medullary portion. J Clin Invest 64:495, 1979
113. Jabs K, Zeidel M, Silva P: Prostaglandin E_2 inhibits Na^+-K^+-ATPase activity in the inner medullary collecting duct. Am J Physiol 257:F424, 1989
114. Lefkowith J: Essential fatty acid deficiency inhibits the in vivo generation of leukotriene B_4 and suppresses levels of resident and elicited leukocytes in acute inflammation. J Immunol 140:228, 1988
115. Rovin B, Lefkowith J, Schreiner GF: Mechanisms underlying the anti-inflammatory effects of essential fatty acid deficiency in experimental glo-

merulonephritis: inhibited release of a monocyte chemoattractant by glomeruli. J Immunol 145:1238, 1990
116. Takahaski K, Schreiner G, Yamashita K et al: Predominant functional roles for thromboxane A_2 and prostaglandine E_2 during late nephrotoxic serum glomerulonephritis in the rat. J Clin Invest 85:1974, 1990
117. Rovin B, Lefkowith J, Lange L, Schreiner G: Glomerular production of a lipid monocyte chemotactic factor, abstracted. J Am Soc Nephrol 1:536, 1990
118. Muir A, Rovin B, Lacy P, Schreiner G: Murine islets treated with in vivo streptozotocin release a macrophage specific chemotactic lipid. Diabetes (in press)
119. Nathan CF: Secretory products of macrophages. J Clin Invest 79:319, 1987
120. Rovin BH, Schreiner GF: Cell-mediated immunity in glomerular disease. Annu Rev Med 42:25, 1990

9

Lipoprotein Interactions with Glomerular Cells and Matrix

Joel Neugarten
Detlef Schlondorff

INTRODUCTION

Uptake of lipids is essential for the survival and growth of all cells. It is to be expected, therefore, that the uptake of lipoproteins is subject to numerous regulatory mechanisms. This is especially true for low-density lipoprotein (LDL) and modified forms of LDL. To a large extent, uptake of LDL and modified LDL is determined by the presence or absence of specific receptors on cell surfaces, the LDL and scavenger LDL receptors. It should be noted, however, that receptor-independent uptake of LDL can also be considerable.

In principle, cellular uptake of lipoproteins will depend in part on the concentration of the lipoprotein to which the cell is exposed. For example, LDL particles may have limited access to perivascular cells because most endothelial beds and their basement membranes are relatively impermeable to particles the size of LDL. In the renal glomerulus, however, the mesangium is separated from the capillary lumen only by a fenestrated endothelium without an intervening basement membrane[1] and thus is probably constantly exposed to LDL. These considerations, together with observations that lipids accumulate within the mesangium in renal diseases and in certain hyperlipidemias, have prompted investigation of the handling of LDL by the mesangial cell. Similar considerations may also apply to the renal interstitium, as peritubular capillaries and ascending vasa recta are also lined by a fenestrated endothelium that is highly permeable to macromolecules. Moreover, accumulation of lipid-laden foam cells in the renal interstitium is a common finding in renal diseases with secondary hyperlipidemia. These observations may be of particular interest, as the severity of interstitial disease appears to be an important prognostic indicator for the progression of renal diseases. Thus, pathologic changes in the glomerulus and interstitium may in part be linked by a common etiologic factor, namely, local lipid accumulation. In the present overview, we summarize the interactions of LDL with cells and matrix in general and with the mesangium in particular, and the consequences that these interactions may have for mesangial biology.

GENERAL BACKGROUND

Receptor-Mediated Uptake of Low-Density Lipoprotein

The majority (two-thirds) of the in vivo clearance of LDL is mediated by cell surface receptors that recognize apolipoprotein B.[2,3] The remainder is accomplished by receptor-independent endocytosis. Receptor-mediated uptake of LDL requires positively charged lysine amino groups of apolipoprotein B_{100} on the LDL particle.[4,5] In addition, apolipoprotein E, present on intermediate-density lipoprotein and β-very low-density lipoprotein (VLDL), is recognized by the LDL receptor. LDL receptors are clustered within clathrin-coated pits on the cell surface[6] (Fig. 9-1). After binding LDL, the pits invaginate and pinch off to form endocytic vesicles that shed their clathrin coat and fuse with one another to form endosomes.[6] LDL receptors dissociate from LDL due to a decline in pH within the endosome.[6] The unoccupied LDL receptor returns to the surface of the cell and thus recycles once every 10 minutes. After receptor dissociation, LDL is delivered to lysosomes where the lipids are hydrolysed to release free cholesterol.[6]

In response to intracellular cholesterol accumulation, LDL receptors are

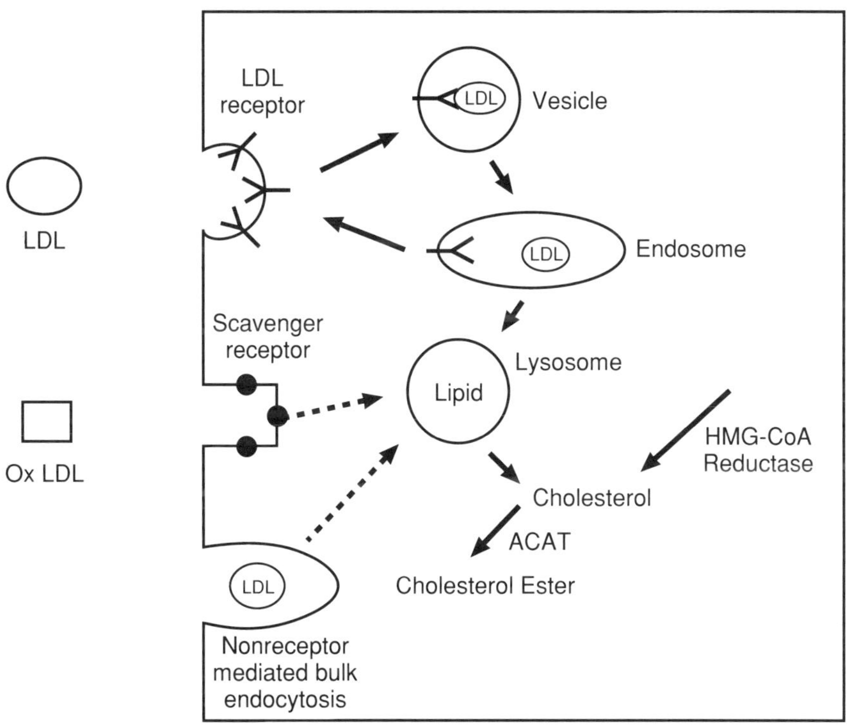

Fig. 9-1. Schematic representation of cellular uptake of LDL and oxidatively modified LDL (Ox-LDL), and intracellular cholesterol metabolism.

downregulated by suppression of transcription for receptor mRNA.[6] This prevents further uptake and accumulation of intracellular cholesterol.[6] Cholesterol is used for membrane synthesis, and excess cholesterol is esterfied by acyl-CoA cholesterol acyltransferase (ACAT). This results in esterfication of cholesterol released from LDL into intracellular cholesteryl esters, predominantly cholesteryl oleate. In addition, de novo cholesterol synthesis via the β-hydroxy-β-methylglutaryl (HMG)-CoA reductase pathway is reduced to further decrease intracellular cholesterol.[6] This occurs as a result of suppression of transcription of the HMG-CoA reductase gene and accelerated degradation of the enzyme protein.[6] Thus, intracellular cholesterol levels regulate both receptor-mediated uptake and de novo synthesis of cholesterol.

Both receptor-dependent and independent uptake of LDL result in appropriate metabolic responses that prevent intracellular lipid accumulation and foam cell formation. It therefore appears impossible to generate foam cells by providing excess LDL to cells, as long as the LDL remains in its native (unmodified) form.

Receptor-Mediated Uptake of Modified Low-Density Lipoprotein

Chemical modification of LDL reduces or abolishes its recognition by the LDL receptor and enhances uptake by a class of scavenger receptors that preferentially bind, internalize, and degrade modified LDL.[4] Scavenger receptors have been demonstrated on a variety of cells, including peritoneal and glomerular macrophages, hepatic Kupffer and sinusoidal cells, and vascular endothelial and smooth muscle cells.[7] Modification of both the apoprotein and lipid moieties of LDL contribute to its altered metabolism.[8] Activated cells generate lipid peroxides and release chemically reactive aldehydes capable of interacting with LDL.[9] The aldehydes, such as malondialdehyde, covalently modify lysine residues of apoprotein B, resulting in net loss of positive charge and altered receptor recognition.[7] This "oxidized" LDL also contains lipid peroxides and partially hydrolysed phospholipids, such as lysophosphatidylcholine.[9,10] Oxidized LDL is characterized by increases in density and in negative charge.[9] It can be identified by increased electrophoretic mobility and by bioassay because of its rapid uptake by macrophages via the scavenger receptors and its toxicity to proliferating cells.[10]

Macrophages appear to have different classes of scavenger receptors that preferentially bind and take up either acetylated or oxidized LDL.[11] The different LDL scavenger receptors have been characterized predominantly by binding studies. Recently, however, two closely related types of scavenger receptors for LDL have been purified and cloned.[12,13] They belong to a new class of receptors with no similarity to the LDL receptor. The two cloned receptors differ only in their C-terminal extracellular domains and may result from alternative splicing of a single gene product. Both receptors have a common collagenous extracellular domain, which appears to be important for binding of modified LDL.[11,13] The collagenous character of the binding site on the LDL scavenger receptor may be of special interest as modified LDL tends to bind to matrix components, including collagens (see below). Little is known concerning regulation of the scavenger receptors. The phorbol ester PMA, an activator of protein kinase C, markedly enhances mRNA levels for the type I scavenger receptor and increases uptake of modified LDL by human monocyte cell lines.[14,15] No studies have been reported on the effects of uptake of modified LDL on mRNA levels for the scavenger receptors. However, in contrast to LDL, uptake of modified LDL by the scavenger receptor is not downregulated by intracellular cholesterol accumulation. Thus, uptake of modified LDL can lead to massive accumulation of free and esterified cholesterol within macrophages, and possibly within vascular smooth muscle cells, resulting in foam cell formation.[4]

In vitro, numerous cell types are capable of modifying LDL to forms recognized by the scavenger receptor.[16] These include macrophages, monocytes, neutrophils, aggregating platelets, vascular endothelial cells, vascular smooth muscle cells,[16] and mesangial cells.[17] LDL also undergoes autooxidation, which is accelerated by incubation with transition metals in cell-free medium.[10] In vivo oxidation of LDL is restricted to the interstitial microenvironment insofar as circulating antioxidants and HDL prevent oxidation within the plasma.[9] As the mesangium is constantly percolated by plasma, LDL localized to the mesangium should be protected from oxidative modification under physiologic circumstances. However, this situation may change in inflammatory glomerular diseases that are characterized by enhanced generation of reactive oxygen species capable of oxidizing LDL.[18]

The source and precise role of reactive oxygen species in modification of LDL remain controversial. Activated cells, including cultured arterial smooth muscle cells, neutrophils, monocytes, and macrophages, can generate reactive oxygen species, such as superoxide anion and hydrogen peroxide, which are capable of oxidizing LDL.[19,20] Moreover, the free radical scavengers vitamin E and glutathione, and divalent cation chelators, such as ethylenediaminetetraacetic acid (EDTA), inhibit cell-mediated oxidation of LDL.[21] Inconsistent results obtained with superoxide dismutase, a superoxide anion scavenger, may be due to varied experimental conditions.[9] Cyclooxygenase-mediated LDL oxidation appears unlikely, but an important role for the lipoxygenase pathway has been recently proposed.[22] Probucol, a potent antioxidant transported in plasma primarily within the lipoprotein fraction, inhibits in vitro oxidation of LDL.[23] In addition, LDL isolated from animals or patients receiving probucol is resistant to subsequent in vitro oxidation.[24]

It has been recently hypothesized that oxidized rather than native LDL plays an essential role in the development of atherosclerotic vascular lesions.[9] Evidence for accumulation of oxidatively modified LDL in atherosclerotic lesions comes from studies demonstrating the following:

1. The presence of modified apolipoprotein B-containing lipoprotein particles in extracts from the lesions[25]

2. Recognition of lipoproteins extracted from the lesions by scavenger receptors on macrophages[26,27]

3. Immunohistologic studies demonstrating modified LDL in the subintimal lesions using antibodies raised against modified LDL[27–29]

Moreover, foam cells, which are essential participants in the development of atherosclerotic lesions, cannot be generated by incubating cells with unmodified LDL, but result from incubating macrophages with the following:

1. Modified LDL extracted from fatty streaks or atherosclerotic lesions[26,27]
2. In vitro oxidized LDL or LDL reacted with aldehydes such as malondialdehyde or 4-hedroxynonenal[30,31]
3. LDL bound to matrix components, such as collagen or glycosaminoglycans (see below)
4. In vitro chemically modified LDL, including acetylated LDL[7]

INTERACTIONS OF LOW-DENSITY LIPOPROTEIN AND THE MESANGIUM

Receptor-Mediated Uptake of Low-Density Lipoprotein by Mesangial Cells

Several recent studies have provided evidence for the presence of specific LDL receptors on mesangial cells cultured from isolated rat and human glomeruli.[32,33] The evidence is derived from binding and uptake studies using radiolabeled LDL, from studies of LDL-induced ^{14}C-labeled oleate incorporation into cholesteryl esters, and from molecular biology studies demonstrating mRNA for the LDL receptor in mesangial cells. Wasserman et al.[32] studied binding of radiolabeled human LDL to cultured rat mesangial cells. Equilibrium binding experiments were performed by increasing the concentration of unlabeled lipoprotein in the presence of a constant amount of the labeled moiety. LDL binding to cell membrane receptors was saturable, with 75 percent displacement occurring between 100 and 200 μg/ml of unlabeled LDL and half-maximal displacement at 5 μg/ml (Fig. 9-2). An estimated 1,000 LDL particles were bound per mesangial cell. The function of LDL binding sites on cultured mesangial cells was demonstrated by uptake studies with ^{125}I- and fluorescent-labeled LDL. As expected, LDL uptake considerably exceeded binding and continued over hours, consistent with recycling of LDL receptors (Fig. 9-3).

In general, cellular uptake of LDL provides cells with cholesterol and should enhance incorporation of fatty acids, such as ^{14}C-labeled oleate, into cholesteryl esters.[6] Consistent with this formulation, incubation of mesangial cells with ^{14}C-labeled oleate and increasing concentrations of LDL led to enhanced ^{14}C-labeled oleate incorporation into cholesteryl esters[34] (Fig. 9-4). Finally, uptake of LDL by cultured mesangial cells was confirmed by using the fluorescent LDL derivative, 1,1′-dioctadecyl, 3,3,3′3′ tetramethylin-

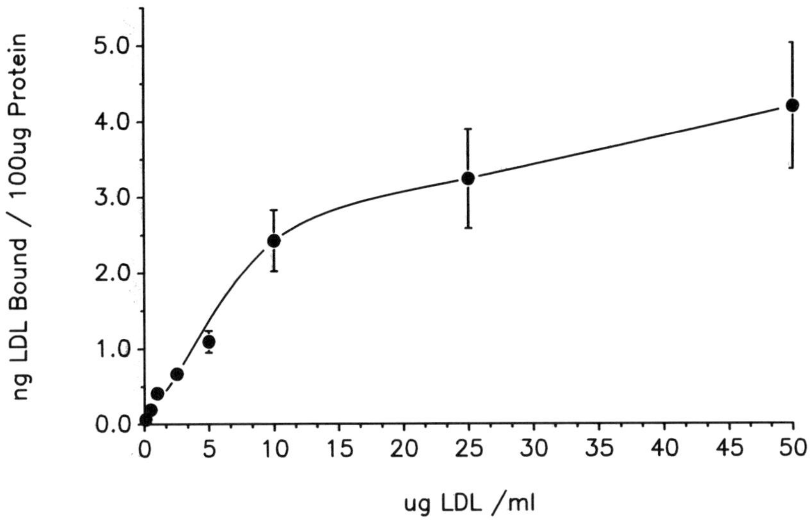

Fig. 9-2. Concentration-dependent binding of LDL to cultured mesangial cells. (From Wasserman et al.,[32] with permission.)

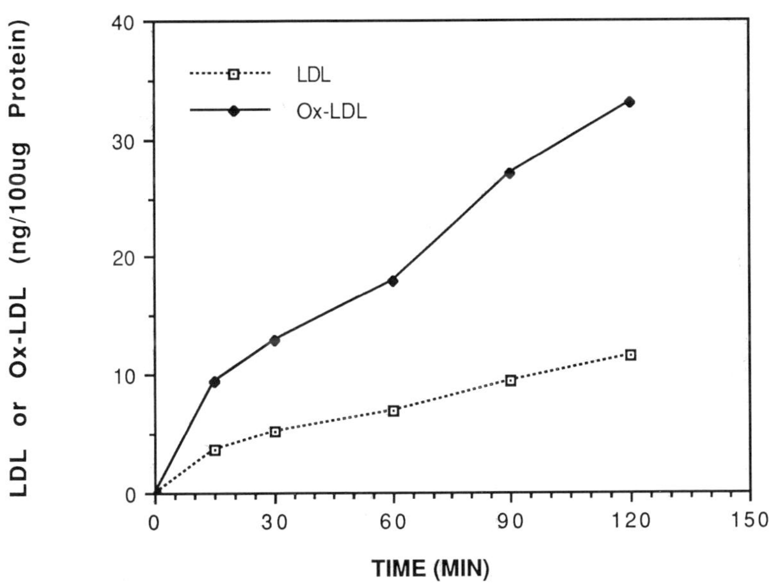

Fig. 9-3. Time-dependent uptake of ^{125}I-labeled LDL or oxidatively modified LDL (Ox-LDL) by cultured mesangial cells. (From Coritsides et al.,[34] with permission.)

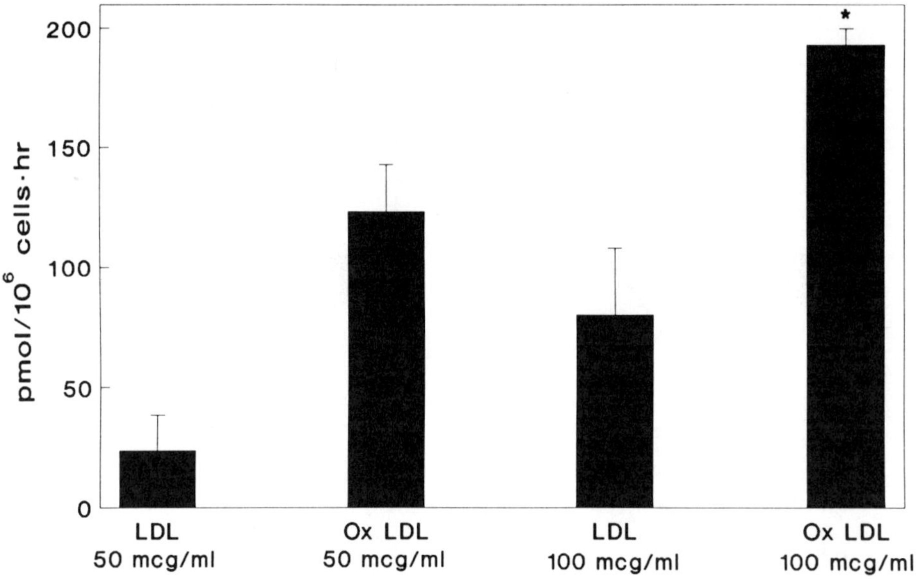

Fig. 9-4. Effects of coincubation of cultured mesangial cells with different concentrations of LDL or oxidatively modified LDL (Ox-LDL) on ^{14}C-labeled oleate incorporation into cholesteryl esters. (From Coritsides et al.,[34] with permission.)

docarbocyanide LDL. Fluorescence microscopy showed that all cells accumulated fluorescent LDL in a pattern consistent with lysosomal accumulation.[32] Although these studies were carried out with human LDL and cultured rat mesangial cells (i.e., across species), comparable results have been obtained with human LDL and human mesangial cells.[33,35]

Preincubation of mesangial cells with increasing concentrations of LDL decreased de novo cholesterol synthesis (Gupta et al., unpublished observation). Thus, cultured mesangial cells regulate their cholesterol synthesis in a manner that should prevent cellular cholesterol accumulation, as described for other cells. These experimental data clearly support the existence of regulated LDL uptake by cultured mesangial cells. To confirm that uptake occurred via LDL receptors, expression of mRNA for the LDL receptor was examined in cultured mesangial cells. Complementary molecular probes for the rat LDL receptor[36] provided unequivocal evidence for the expression of mRNA for LDL receptors in cultured mouse and rat mesangial cells (Fig. 9-5; Schlondorff et al., unpublished observation).

Receptor-mediated binding, internalization, and degradation of lipoproteins by cultured human glomerular epithelial cells have also been reported.[33,37] Uptake of apo B,E-containing lipoproteins, such as chylomicron remnants and β-VLDL, was enhanced compared with uptake of apo B_{100}-containing LDL.[37] In addition, receptor-mediated lipoprotein uptake suppressed cellular synthesis of cholesterol.[37]

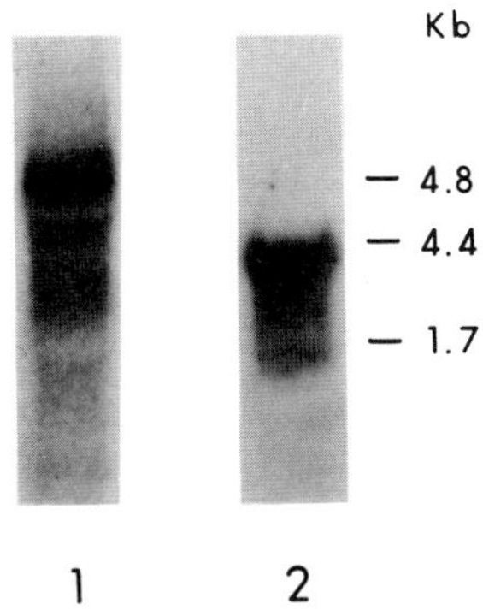

Fig. 9-5. Northern blot analysis for expression of mRNA for the LDL and scavenger LDL receptor. Twenty micrograms of total RNA extracted from rat mesangial cells were hybridized with either a complementary probe for the rat LDL receptor in lane 1[36] or in lane 2 with a probe for the bovine type II scavenger LDL receptor.[12] (Schlondorff et al., unpublished observation.)

Uptake of Modified Low-Density Lipoprotein by Scavenger Receptors on Mesangial Cells

Mesangial cells not only express receptors for native LDL, but may also have scavenger receptors for modified and, specifically, oxidized LDL. Therefore, experiments were performed to compare binding and uptake of native human LDL with that of LDL oxidized in vitro by incubation with copper.[34] Oxidized LDL bound to cultured rat mesangial cells in a concentration-dependent manner. Binding of oxidized LDL exceeded that of native LDL approximately four- to fivefold. The binding affinity for oxidized LDL was lower than that for native LDL. In addition, considerable amounts of oxidized LDL bound to mesangial cells in a nonspecific manner and could not be fully displaced by high concentrations of oxidized LDL.

These results suggested the existence of specific binding of oxidized LDL with a lower affinity than native LDL but with a somewhat large degree of nonspecific binding. The specificity of binding was examined by competitive binding. The specificity of binding was examined by competitive displacement of oxidized LDL by native LDL, and vice versa (Fig. 9-6). One hundred μg/ml of oxidized LDL was able to displace LDL binding to approximately the same extent (75 percent) as an equal concentration of LDL. In contrast, 100 μg/ml of native LDL displaced oxidized LDL only by 15 percent, suggesting that native LDL bound only to the LDL receptor and not to binding sites for oxidized LDL.

The effects of heparin and polyinosinic acid on binding of LDL and oxidized LDL to mesangial cells were also examined.[34] Heparin inhibits binding of native LDL to its receptor in numerous cell types.[38] Heparin interacts ionically with the positively changed lysine residues of LDL to inhibit binding of LDL to the LDL receptor. In addition, heparin and certain polyanions, such

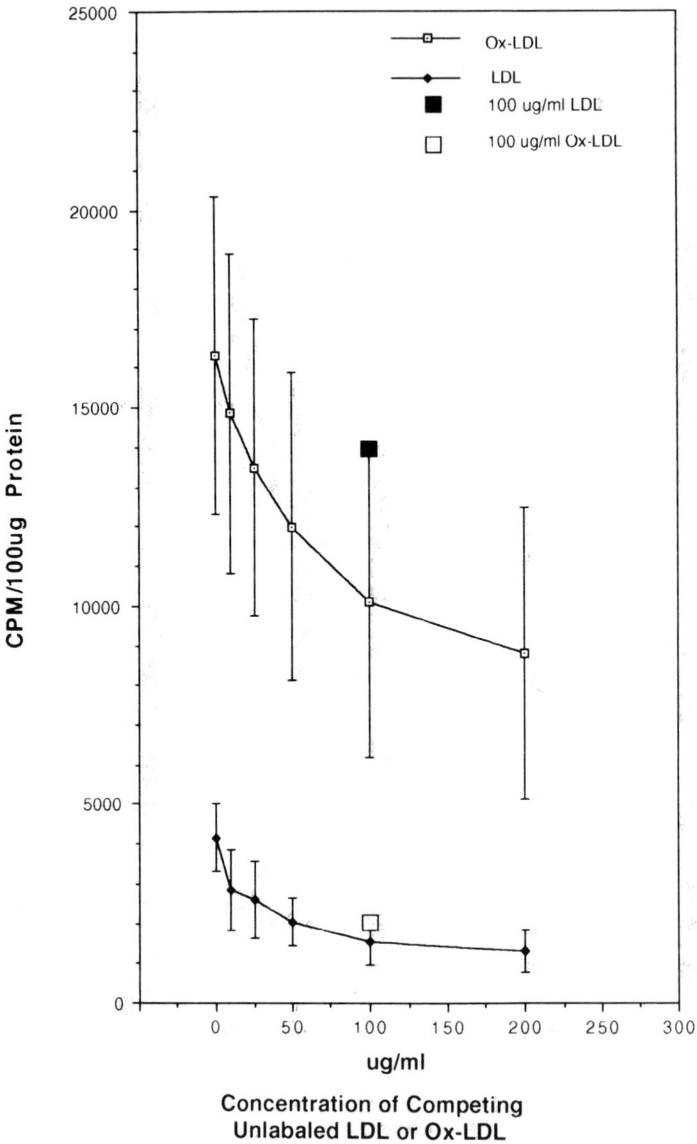

Fig. 9-6. Displacement of ^{125}I-labeled LDL or oxidatively modified LDL (Ox-LDL) bound to mesangial cells by increasing concentrations of unlabeled LDL or Ox-LDL. The large closed square indicates displacement of ^{125}I-labeled Ox-LDL by unlabeled LDL and the large open square indicates displacement of ^{125}I-labeled LDL by unlabeled Ox-LDL, respectively. From Coritsides et al.,[34] with permission.)

as polyinosinic acid, block binding of oxidized LDL to the scavenger receptor.[34] Polyinosinic acid should not, however, interfere with LDL binding to its receptor. Heparin decreased both LDL and oxidized LDL binding to mesangial cells by 50 to 70 percent[34] (Fig. 9-7). In contrast, polyinosinic acid inhibited binding of oxidized LDL, but not LDL. These findings confirmed different binding sites for LDL and oxidized LDL on mesangial cells. Characteristics of the receptors resemble those reported for macrophage receptors,[9] indicating that mesangial cells in culture express both LDL receptors and a scavenger receptor for oxidized LDL.

To confirm uptake of oxidized LDL by mesangial scavenger receptors, uptake of oxidized LDL and incorporation of oleic acid into cholesteryl oleate were determined in the presence of oxidized and native LDL.[34] Mesangial cell uptake of oxidized LDL exceeded that of native LDL by three- to fourfold and failed to plateau during a 2-hour incubation (Fig. 9-3). A substantially greater increase in intracellular accumulation of cholesteryl oleate was observed when mesangial cells were incubated with oxidized compared with native LDL, indicating the presence of mesangial scavenger receptors and activation of ACAT by the accumulation of cholesterol in the mesangial cell (Fig. 9-4). To confirm that uptake occurred via scavenger LDL receptors, expression of mRNA for the scavenger receptor was examined in cultured mesangial cells. Complementary molecular probes for the bovine type II scavenger LDL receptor[13] provided unequivocal evidence for the expression of mRNA for scavenger LDL receptors in cultured rat and mouse mesangial cells (Fig. 9-5; Schlondorff et al., unpublished observation).

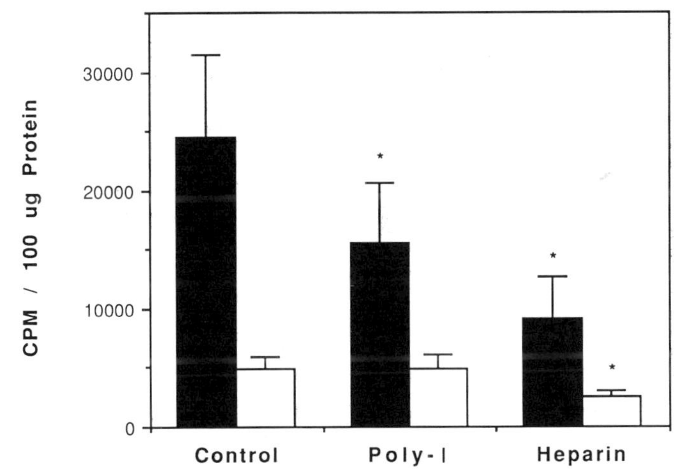

Fig. 9-7. Effect of heparin and polyinosinic acid on the binding of ^{125}I-labeled LDL (open bars) and ^{125}I-labeled oxidatively modified LDL (Ox-LDL) (closed bars) to mesangial cells. Binding of ^{125}I-labeled LDL or Ox-LDL in the presence or absence of heparin sulfate (10 mg/ml) or polyinosinic acid (20 μg/ml) was evaluated after 2 hours of incubation at 4°C. Asterisks indicate $P < .05$ or better compared with the respective controls. (From Coritsides et al.,[34] with permission.)

Uptake of Glycated Low-Density Lipoprotein by Mesangial Cells

Apart from oxidative modification, LDL can also be modified by glycation due to high glucose concentrations. Sera obtained from patients and experimental animals with diabetes contain elevated levels of glycated LDL compared with normoglycemic[39,40] controls, analogous to elevated levels of glycated hemoglobin A_{1c}. Glucose covalently binds to ϵ amino groups on lysine residues of apolipoprotein B.[39,40] Glycation of LDL impairs receptor mediated uptake and degradation of LDL by cultured fibroblasts.[39,41–43] This impairment has been correlated with the degree of LDL glycation.[39,42,44] In most studies, recognition by the LDL receptor was reduced when 2.5 to 8 percent of the lysine residues were altered, as could occur in poorly controlled diabetic patients.[39,42,44] The in vivo fractional catabolic rate for glycated LDL was reduced as well.[39] One group claimed, however, that reduced binding is observed only at unphysiologically high degrees (>20 percent) of LDL glycation.[45]

We compared binding and uptake of native LDL and glycated LDL by cultured rat mesangial cells. Moderate degrees of glycation, as can occur in diabetic patients (5 to 8 percent), were induced in vitro by incubating LDL with 200 mmol/L glucose. Glycation of LDL reduced mesangial cell binding and uptake by approximately 15 to 20 percent compared with control LDL (Gupta and Schlondorff, unpublished observation). The reduced uptake was also reflected by a lower degree of ^{14}C-labeled oleate incorporation into cellular cholesteryl esters. In these experiments, special attention was paid to prevent oxidative modification of the LDL during the 1 to 2 weeks of incubation with glucose required to achieve the appropriate degree of glycation. Furthermore, we found no evidence that glycated LDL was cytotoxic for mesangial cells. Taken together, our experiments confirm the conclusion drawn by most investigators studying other cell types, that glycation of LDL to a degree observed in diabetes mellitus (i.e., 3 to 8 percent) modestly inhibits LDL uptake.

The decrease in receptor-mediated uptake of glycated LDL in vivo may increase the amount of LDL that remains in the mesangial matrix of the diabetic patient, where it may bind to the expanded and altered matrix components of the diabetic mesangium and undergo further modification, such as oxidation. These events may contribute to progressive glomerulosclerosis in diabetic nephropathy.

Effect of Vasoactive Agents on Uptake of Low-Density Lipoprotein by Mesangial Cells

Vasoactive agents such as angiotensin and eicosanoids could influence mesangial delivery of LDL in vivo by altering filtration factors. We recently found that angiotensin-independent hemodynamic forces could directly enhance macromolecular uptake by cultured mesangial cells (Singhal and Schlondorff, unpublished observation). This effect appeared to be mediated by changes in cytoskeletal activity and could be abrogated by agents that increase intracellular cyclic adenosine monophosphate (cAMP) or cyclic guanosine monophosphate (cGMP) levels, such as dopamine or atriopeptin. We also examined the effects of angiotensin II, the thromboxane analog U46619, and prostaglandin E_2 (PGE_2) on LDL uptake by cultured mesangial cells during short-term (up to 1 hour) incubation. Angiotensin and the thromboxane analog enhanced uptake of LDL by mesangial cells, whereas PGE_2 decreased basal as well as angiotensin-induced LDL uptake. Furthermore, inhibition of endogenous PGE_2 formation by meclofenamate enhanced the effects of angiotensin II and the thromboxane analog on LDL uptake. Thus, it appears that angiotensin II and thromboxane can increase LDL uptake, an effect opposed in a feedback manner by PGE_2. These observations may be of interest for mesangial handling of LDL because increased levels of angiotensin and thromboxane occur in many forms of progressive renal diseases, and perhaps in hypercholesterolemia in general.[46]

Additional effects of vasoactive agents on LDL metabolism may exist. Walli et al.[35] recently reported that pretreatment of mesangial cells with endothelin doubled their uptake of LDL. In contrast to the effects of angiotensin described above, this was not an acute event, but required 15 hours of preincubation with endothelin. There was, however, no change in the number of LDL receptors after endothelin treatment, suggesting the possibility that vasoactive agents may influence the endocytic pathway for LDL or recycling of receptors.

In Vivo Uptake of Low-Density Lipoprotein and Oxidized Low-Density Lipoprotein by Glomeruli

To examine whether oxidized LDL is taken up in preference to native LDL by glomeruli in vivo, we injected radiolabeled LDL and oxidized LDL into the suprarenal aorta of rats.[34] Radioactivity was then determined in isolated glomeruli and tubular suspensions and was compared with uptake by the spleen and liver. Our studies demonstrated preferential mesangial cell binding and uptake of oxidized LDL compared with native LDL (Fig. 9-8). Consistent with published results, native LDL was taken up predominantly by the liver and also by the spleen.[47] Glomerular uptake of native LDL was greater than that by tubular tissue, but was only 20 percent of that observed in liver when expressed per milligram of tissue protein. Uptake of oxidized LDL markedly exceeded that of LDL in all tissues examined. The increased uptake of labeled oxidized LDL was particularly striking in the glomeruli, so that glomerular content of oxidized LDL per milligram of tissue protein reached levels observed in liver.

To confirm the glomerular localization of oxidized LDL, we used autoradiography on renal cortical slices. Injection of ^{125}I-labeled LDL resulted in insufficient radioactivity to allow detection by autoradiography. In contrast, after infusion of ^{125}I-labeled oxidized LDL, multiple grains could be localized to each glomerulus and, to a lesser degree, to the tubules as well. The distribution of the radioactive grains within the glomerulus predominantly

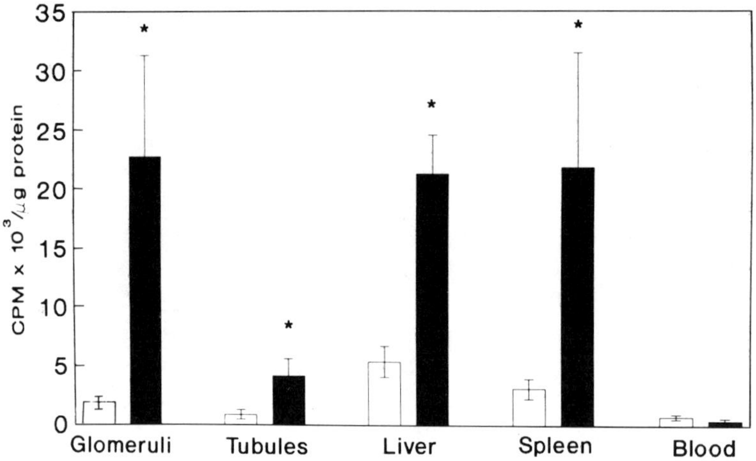

Fig. 9-8. Uptake of LDL compared with oxidatively modified LDL (Ox-LDL) in normal rats in vivo. ^{125}I-labeled LDL or Ox-LDL was infused into the aorta above the renal artery for 10 minutes. Organs and glomeruli were then isolated, and tissue content of ^{125}I-labeled LDL or Ox-LDL was determined and expressed as counts per minute per microgram of tissue protein. Asterisks indicate $P < .05$ or better, comparing content of ^{125}I-labeled LDL versus ^{125}I-labeled Ox-LDL. (From Coritsides et al.,[34] with permission.)

followed a central stalk pattern; however, we were unable to further assign the radioactivity to specific cell types.

PROLIFERATIVE EFFECTS OF LIPOPROTEINS

Numerous investigators have suggested that LDL exerts a proliferative effect or that LDL may act as a coproliferative agent to enhance the effects of cytokines. Arterial explants and cultured vascular smooth muscle cells derived from cholesterol-fed animals show enhanced DNA synthesis and cell proliferation.[48–50] In addition, hyperlipidemic serum has been shown to enhance proliferation of cultured fibroblasts, endothelial and smooth muscle cells, and aortic explants in a variety of species.[48–58] The mitogenic activity has been localized to the LDL fraction in most of these studies.[51,53,56,57,59,60] Other investigators, however, have found that hyperlipidemic and control sera stimulate cell proliferation to a similar extent.[56,59,61,62] In some of these latter studies, mitogenic activity was localized to the LDL fraction of normal serum,[61] while in others a nonlipoprotein mitogenic serum factor was isolated.[59,62] The proliferative effect of LDL on cultured cells may merely represent an in vitro phenomenon, reflecting only the lipid requirement of proliferating cells in culture. For example, the proliferative effect of lipoproteins on mitogen-activated T cells can be mimicked by crude soybean lipid preparations or similar lipid mixtures.[63]

Several studies have shown a biphasic growth response to LDL in cultured cells: low concentrations of LDL stimulate proliferation of vascular endothelial and smooth muscle cells, whereas higher concentrations suppress cell proliferation and induce cell injury.[64,65] However, the inhibitory effect of high concentrations of LDL may be due to oxidation of LDL during the experimental incubation (see below).

CYTOTOXIC EFFECTS OF LIPOPROTEINS

Various lipoprotein fractions are cytotoxic to cells in culture.[21,66–73] Numerous investigators have demonstrated that it is the oxidized forms of LDL and VLDL, and not the native moieties, that are responsible for cytotoxicity. If lipoprotein fractions are prepared in the presence of free radical scavengers to prevent in vitro autooxidation during the fractionation, dialysis, and storage of the lipoprotein fractions, cytotoxicity is ameliorated or eliminated.[21,66,70,74] Antioxidants such as butylated hydroxytoluene, vitamin E, or EDTA, when added during lipoprotein preparation, abolished LDL or VLDL-induced cytotoxicity, whereas superoxide dismutase and catalase partially ameliorated cytotoxicity.[21,66,70,71,75] The addition of antioxidants to the medium after incubation of cells with lipoproteins failed to reverse cytotoxicity, indicating that cytotoxicity is mediated by products of lipid oxidation and not by free radicals generated in the medium.[21] High-density lipoprotein (HDL) can also protect against cell injury during LDL oxidation, most likely by acting as a scavenger for free radicals or by taking up toxic lipid products.[68] The cytotoxic component of oxidized LDL is present in the lipid portion,[21,71,76] most likely representing lysophospholipids.[77] Cytotoxicity correlates with the extent of formation of thiobarbituric acid-reactive substances (TBARS), which also reflects oxidative modification of the lipids.[66,71,76]

The susceptibility of cells to the toxic effects of oxidized LDL is related to a variety of factors, including cell density, proliferative activity, and phase of the cell cycle. Studies performed in synchronized fibroblast cultures have demonstrated that oxidized LDL induces cell injury and death toward the end of the S phase of the cell cycle.[78] The addition of hydroxyurea to the medium, by preventing cells from entering the DNA synthetic phase, prevented oxidized LDL-induced cell death.[78] When hydroxyurea was added to synchronized cells after completion of the S phase, oxidized LDL failed to injure cells in subsequent phases.[78] In addition, susceptibility to LDL-induced injury is reduced in slowly proliferating cultures of high cell density.[69,73,78]

It has been suggested that free radical-mediated lipoprotein oxidation accompanies hyperglycemia and leads to the appearance of lipid peroxidation products in the serum. Indeed, lipoproteins containing lipid peroxides have been demonstrated in the serum in experimental diabetes and in insulin-dependent diabetes in humans.[67,69] Enhanced injury to cultured endothelial cells is observed on incubation with LDL or VLDL obtained from rats with experimental diabetes.[67,69,70] Cytotoxicity is reversed by maintenance of euglycemia with insulin or by treatment of hyperglycemic animals with antioxidants.[67,69,70] Reduced cytotoxicity was correlated with a decline in TBARS in lipoprotein fractions.[67,70] Thus, these studies suggest that hyperglycemia is associated with enhanced generation of lipid peroxides, which is reflected in the greater cytotoxicity of lipoprotein fractions obtained from diabetic sera.

EFFECTS OF LIPOPROTEINS ON MESANGIAL CELL PROLIFERATION AND VIABILITY

Native Low-Density Lipoprotein

The effects of native LDL on cultured rat and human mesangial cells have been studied by ourselves and other groups.[32,34,35] In our studies, cultured rat mesangial cells were incubated with human LDL, whereas other investigators incubated human mesangial cells with human LDL. An increase in ^{3}H-labeled thymidine incorporation into mesangial DNA is observed after the addition of 10 to 50 μg of LDL protein/ml culture medium (Fig. 9-9). These results have been confirmed by determination of cell counts (Fig. 9-9). At higher concentrations of LDL (from 100 to 500 μg/ml), the results vary somewhat: some investigators report continued stimulation of ^{3}H-labeled thymidine uptake, while others report a return to control levels or even an inhibition of thymidine uptake at LDL concentrations of 250 μg/ml. The inhibition of cell proliferation at the high LDL concentrations may be due to oxidative modification of the LDL that occurred during preparation of the LDL or during incubation with the mesangial cells. The modest increase in mesangial cell proliferation with low concentrations of LDL is consistent with the published results in other cell culture systems.

Concentrations of LDL resulting in cell proliferation in culture are in the range of 10 to 50 μg (i.e., no more than 10 percent of normal in vivo plasma LDL concentrations). Given the relatively unimpaired access of plasma to the glomerular mesangium, mesangial cells in vivo are probably always exposed to LDL concentrations that in culture would result in maximal proliferation. Therefore, isolated changes in plasma LDL concentrations are unlikely to influence mesangial cell proliferation in vivo, especially since mesangial cells in vivo have a slow turnover.[79]

While LDL by itself may only slightly enhance cell proliferation, LDL markedly potentiates mesangial cell proliferation in response to endothelin and platelet-derived growth factor,[35,80] known mitogens for mesangial cells.[79] Thus, under certain culture conditions, LDL may represent an important comitogen, perhaps by providing adequate lipid and cholesterol needed for proliferating cells.

Oxidized Low-Density Lipoprotein

Oxidized LDL causes a concentration-dependent decrease in thymidine incorporation and cell number in cultured mesangial cells[32] (Fig. 9-9). This effect of oxidized LDL is already apparent at concentrations of 25 μg/ml.

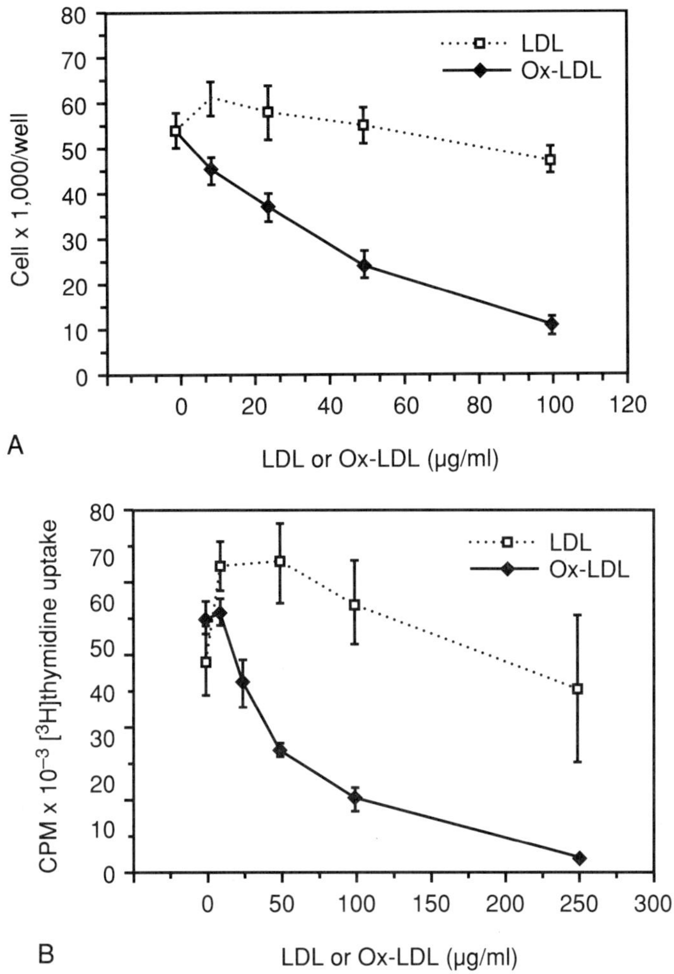

Fig. 9-9. Effect of oxidatively modified LDL (Ox-LDL) or LDL on **(A)** mesangial cell number or **(B)** ^{3}H-labeled thymidine incorporation by mesangial cells. Thymidine incorporation into mesangial cells was determined in the presence of increasing concentrations of LDL or Ox-LDL during 24 hours of culture. Cell number was determined after 72 hours of culture. (From Coritsides et al.,[34] with permission.)

Lysophosphatidycholine may be responsible in large part for the toxic effect of oxidized LDL.[77] Enhanced synthesis of PGE_2 does not contribute to the inhibition of mesangial cell proliferation by oxidized LDL insofar as the decrease in thymidine incorporation persists in the presence of a prostaglandin synthesis inhibitor.[32] Direct observations under phase contrast microscopy also confirmed the detrimental effect of oxidized LDL on mesangial cells.[32]

INTERACTIONS OF LIPOPROTEINS WITH MATRIX COMPONENTS

Lipoprotein–Proteoglycan Interactions

A large body of experimental evidence suggests that the binding of lipoproteins to extracellular matrix components in the arterial wall may contribute to the development of atherosclerotic lesions. Immunofluorescence microscopy and histochemical techniques have localized apolipoprotein B in atherosclerotic lesions, predominantly in an extracellular location, and in close association with elastin and collagen fibers.[81] While most of the uptake of LDL by the arterial wall is due to cellular metabolism, with the intima accounting for 40 percent, a significant proportion of the LDL taken up remains in the extracellular matrix.[82] In fact, the concentration of LDL has been estimated to range from 10 percent of plasma levels in interstitial fluid to as high as 50 percent in the aortic subendothelial space and even higher in advanced atherosclerotic lesions.[82] Only 35 to 40 percent of extracellular lipids in atherosclerotic vessels can be extracted with saline solutions (i.e., are free within the extracellular space or loosely associated with matrix).[83] Most of the remainder may be bound to extracellular mucopolysaccharides, as suggested by its release following sequential treatment with chondroitinase, collagenase, and elastase.[83] Furthermore, intact proteoglycan–lipoprotein complexes have been extracted from fatty streak lesions and fibrous plaques of aortas from atherosclerotic humans and hypercholesterolemic animals.[84–88]

Plasma lipoproteins may insudate the perivascular space and gain access to proteoglycan glycosaminoglycans of the subendothelial matrix. Proteoglycans may impede the movement of LDL through the matrix by several mechanisms. Molecular sieving and steric exclusion are determined by the network formed by proteoglycan–hyaluronic acid aggregates, whereas ionic binding sites for LDL may arise from polyanionic glycosaminoglycan sulfate side chains.[89]

Iverius[89] demonstrated the ionic binding of LDL and VLDL to glycosaminoglycan-substituted agarose gels containing dermatan sulfate, heparin sulfate, and chondroitin-4-sulfate at physiologic pH and ionic strength. HDL and acetylated LDL did not bind. Inhibition of binding after modification of the lipoproteins' free amino groups suggested that ionic binding of positively charged amino groups on apolipoprotein B to negatively charged sulfate groups on glycosaminoglycan side chains mediated this interaction. Moreover, the affinity of glycosaminoglycans for lipoproteins was determined by the charge density of the polysaccharide. Many other investigators have confirmed in vitro formation of both soluble and insoluble complexes between LDL and proteoglycans. The formation of insoluble complexes in vitro requires the presence of calcium, which may first interact with phosphate groups on phospholipids to provide cationic sites for binding of anionic sulfate groups on glycosaminoglycans.[90]

Conformational and structural alterations in arterial proteoglycans may influence lipoprotein–proteoglycan interactions.[84] Binding of LDL to proteoglycan requires intact protein and glycosaminoglycan moieties.[84,91] Conformational and structural alterations in LDL may also influence these interactions. Acetylation of lysine or argine residues of LDL prevents complex formation.[91] Moreover, circulating LDL is characterized by heterogeneity of structure and physical properties.[92] Camejo and colleagues identified LDL subclasses from patients with hypercholesterolemia and ischemic heart disease which complexed more readily with human aortic proteoglycans.[92,93] These highly reactive LDLs are characterized by a high ratio of cholesterol esters to triglycerides, a higher isoelectric point, and less sialic acid content.[92,93] Similarly, lipoproteins from cholesterol-fed rabbits showed enhanced ability to form complexes with proteoglycans.[94] Further complicating our understanding of lipid–proteoglycan interactions are alterations in the physical structure and physiologic behavior of LDL that occur as a result of binding to proteoglycans.[95,96] Thus, multiple factors relating to both LDL and matrix components appear to influence their interactions.

Complexes formed due to interactions between lipoproteins and proteoglycans show enhanced receptor-mediated uptake by macrophages compared with native lipoproteins.[97–102] For example, complexes formed on incubation of human LDL with bovine or porcine chondroitin-6-sulfate–rich arterial proteoglycans stimulated lipoprotein uptake by macrophages and enhanced intracellular accumulation of cholesteryl esters.[97,98] The receptor binding site that recognized these complexes was distinguished from the acetyl-LDL receptor.[99] Increased uptake and enhanced cholesteryl ester accumulation in mouse macrophages has also been demonstrated with LDL complexed with dextran sulfate, collagen, and fibronectin.[98,100–102] In addition, lipid–protein complexes extracted from atherosclerotic plaques are rapidly taken up by mouse peritoneal macrophages and enhance intracellular cholesterol ester accumulation.[26] The receptor recognizing these complexes was also distinguished from the acetyl-LDL receptor.[103]

Atherosclerotic lesions in humans and experimentally induced lesions in animals not only show increased proteoglycan content in areas of injury, but also alterations in the composition of proteoglycans with enrichment of sulfated glycosaminoglycans, especially dermatan sulfate.[85,104] It has been suggested that accumulation of plasma LDL within the arterial extracellular matrix may alter the rate of glycosaminoglycan synthesis and induce changes in glycosaminoglycan composition, and that this modification may play a role in further retention of LDL (see below).

Lipoprotein–Collagen Interactions

A role has been suggested for collagen in slowing the diffusion of LDL through the arterial extracellular matrix and in binding of LDL.[82] LDL co-localizes with collagen in immunofluorescence microscopy studies of fatty streak lesions.[82] The concentration of LDL in atherosclerotic lesions is highest in fibrous plaques, and collagenase releases much of it.[105,106] In an in vitro model, the diffusion of native and oxidized LDL into the interstitial medium of a collagen gel was retarded by the molecular sieving effect exerted by the collagen matrix.[82] In addition, the oxidized form of LDL has been shown to preferentially bind to gels of type 1 collagen.[82]

Alterations in collagen by advanced nonenzymatic glycation occur with normal aging and in an accelerated fashion in tissues of diabetic patients. LDL is covalently bound by advanced nonenzymatic glycation products of collagen formed in vitro.[107] Hyperglycemia-induced advanced glycation end products of collagen may contribute to the development of atherosclerotic lesions by binding to amino groups of LDL to enhance LDL accumulation in the arterial wall. Indeed, the lipoprotein content of diabetic aorta is several-fold greater than that of normal animals.[108]

Effects of Low-Density Lipoprotein on Matrix Synthesis

In vitro studies demonstrate that lipoproteins may influence the synthesis and secretion of arterial proteoglycans.[109] LDL exerts a biphasic effect on proteoglycan production by human umbilical vein endothelial cells, increasing synthesis and secretion at LDL concentrations below 200 μg/ml and suppressing proteoglycan production at higher concentrations. In addition, the proteoglycan formed was enriched in chondroitin-6-sulfate, and the size of the chondroitin sulfate was increased due to a larger glycosaminoglycan side chain. A biphasic effect of LDL on glycosaminoglycan secretion by human fetal arterial smooth muscle cells has also been demonstrated.[64] LDL stimulates glycosaminoglycan synthesis and secretion in human aortic smooth muscle cells and fibroblasts.[64] In addition, monkey and rabbit cultured aortic cells show increased synthesis of sulfated proteoglycans, predominantly dextran sulfate, after exposure to LDL.[110] In contrast, a few studies have shown no change or a decline in proteoglycan synthesis after incubation of human fibroblasts with LDL.[110,111]

In most in vitro studies, hyperlipidemic serum failed to increase or, in fact, decreased collagen synthesis in human, rat, monkey, and rabbit aortic smooth muscle cells and in human fibroblasts.[112–117] One study, however, reported enhanced collagen synthesis on incubation of human aortic smooth muscle cells with LDL.[118]

Discrepant results may perhaps be explained in part by inadvertent oxidation of LDL during preparation or after incubation with cells. In addition, some of the effects of LDL on matrix synthesis may be mediated by effects on cell proliferation. As all LDL concentrations that have stimulatory effects in cell culture are within the physiologic serum range for LDL, it is questionable whether these observations in cell culture apply to the in vivo situation.

Low-Density Lipoprotein–Matrix Interactions in the Glomerulus

The interactions of LDL with matrix components are of specific interest in the context of the glomerulus and the mesangial matrix. Mesangial matrix consists of various collagens and glycosaminoglycans (mostly dermatan sulfate, chondroitin sulfate, and some heparan sulfate), as well as laminin, fibronectin, thrombospondin, etc.[79] Thus, all components that can bind and trap LDL are present in the mesangial matrix. We therefore compared the binding of LDL and oxidized LDL to cultured mesangial cells and to extracted matrix that had been generated by the cells. Matrix was prepared by culturing rat mesangial cells for 2 to 3 weeks in plastic culture wells. The cells were then removed with a cell lysis buffer consisting of ammonium hydroxide (0.25 mmol/L) and the nonionic detergent Triton X (0.1 percent). The remaining matrix consisted predominantly of collagens I and IV, laminin, fibronectin, dermatan, and chrondroitin sulfates, but contained little heparan sulfate.[119] Binding studies using LDL and oxidized LDL were then performed on intact mesangial cells or on the matrix they had generated. Binding of both native LDL and oxidized LDL to the matrix markedly exceeded that to intact cells (Fig. 9-10). Binding of LDL and of oxidized LDL to matrix were comparable. In contrast, the binding of oxidized LDL to mesangial cells exceeded that of native LDL. The degree of binding of LDL or oxidized LDL to mesangial cells was saturable, but that of both LDL and oxidized LDL to matrix continued to increase even at concentrations of 100 μg/ml (Fig. 9-10). Thus, the mesangial matrix has a large binding capacity for LDL and oxidized LDL. The exact matrix components responsible for this binding remain to be determined; however, based on the studies mentioned above, likely candidates include collagen and glycosaminoglycans. The interactions of LDL with mesangial matrix may be of interest in the context of the pathophysiologic contribution of hyperlipidemia to glomerulosclerosis, which will be discussed below.

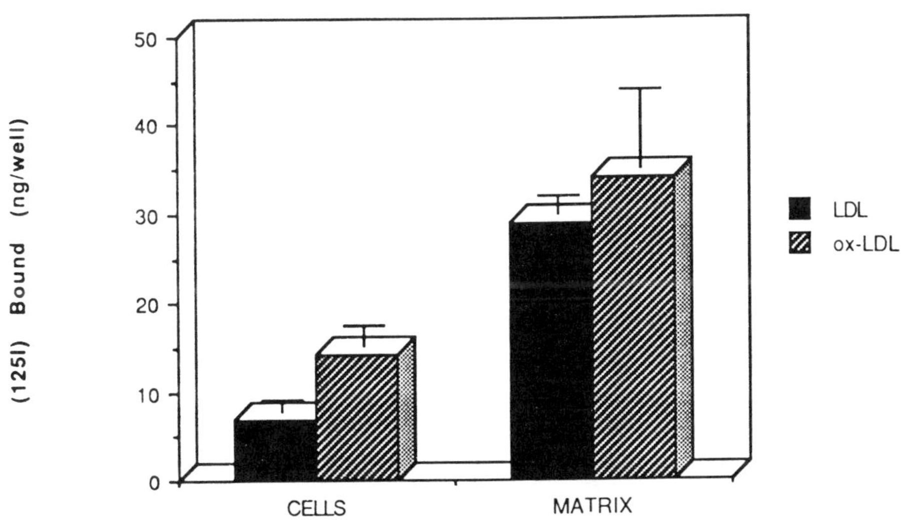

Fig. 9-10. Binding of ^{125}I-labeled LDL or oxidatively modified LDL (Ox-LDL; each at 2 μg/ml) to mesangial cells or to matrix remaining after removal of cells by lysis buffer. (Gupta and Schlondorff, unpublished observation.)

EFFECTS OF LIPOPROTEINS ON PROSTANOID AND CYTOKINE SYNTHESIS

Lipoproteins modulate synthesis and secretion of prostanoids and cytokines by a variety of cell types. Incubation with HDL increased release of prostacyclin and PGE_2 from cultured rabbit aorta smooth muscle and endothelial cells in a dose-dependent manner.[120] LDL had a similar, but less pronounced, effect.[120] Based on observations that labeled arachidonate from lipoproteins is incorporated into newly released prostaglandins and that delipidated apoprotein has no stimulatory effect, it has been proposed that lipoproteins stimulate prostaglandin synthesis by providing the substrate arachidonate to an intracellular pool linked to cyclooxygenase.[120,121] In contrast, other investigators have shown a suppressive effect of LDL on the synthesis of prostacyclin in a variety of tissues.[122] These discrepant results may be reconciled by a biphasic effect of LDL on prostacyclin synthetase activity: stimulatory at low concentrations and inhibitory at high concentrations.[123]

In mouse peritoneal macrophages, oxidized LDL stimulates synthesis of PGE_2, PGI_2, and LTC_4.[124] Native LDL, acetylated LDL, and β-VLDL were without effect. It was suggested that lipid peroxide–lipoprotein complexes increase arachidonate metabolism by stimulating the activity of phospholipase A_2 and/or cyclooxygenase.[124] This would explain the rapid stimulation of prostanoid synthesis in response to oxidized LDL, which contrasts with the slow response to HDL; this last finding appears to be substrate related.

Thromboxane synthesis may also be modulated by lipids. Experimental dietary hypercholesterolemia and familial type II hyperlipidemia in humans are associated with increased platelet thromboxane production.[46] In cholesterol-fed rats, thromboxane and PGE_2 are increased in urine and in proximal tubular fluid; an effect that is prevented by probucol treatment.[46] In addition, aggregation of cholesterol-loaded platelets results in increased release of arachadonic acid from platelet phospholipids and augmented conversion to thromboxane.[125]

Lipoproteins also influence synthesis and secretion of proteases, cytokines, and other agents involved in inflammation and cell growth. Incubation of rat peritoneal macrophages with acetyl LDL or malondialdehyde-modified LDL (oxidized LDL) induces secretion of elastase, cathepsin D, and α-mannosidase.[126] Enhanced synthesis of the monocyte colony-stimulating factor (CSF-1) is seen when cultured endothelial cells are incubated with minimally modified LDL generated by exposure of LDL to iron.[127] Surprisingly, this stimulation appeared to relate to an interaction between minimally modified LDL and the LDL receptor, and not the scavenger LDL receptor. These findings, if confirmed, would be of great interest, as CSF-1 is chemotactic for monocyte macrophages and is required for their survival, proliferation, and differentiation.[128] Increased mRNA level for platelet-derived growth factor has recently been reported in human cultured mesangial cells exposed to LDL or β-VLDL and was linked to a proliferative response.[80] On the other hand, uptake of acetyl LDL and accumulation of intracellular cholesteryl esters by rat peritoneal macrophages suppresses lipopolysaccharide-induced interleukin-1 production.[126]

EFFECTS OF LIPOPROTEINS ON MESANGIAL CELL PROSTAGLANDIN SYNTHESIS

We recently compared the effects of oxidized and native LDL on the synthesis of PGE_2 by rat mesangial cells.[34] LDL moderately increased PGE_2 production by mesangial cells in culture over a 24-hour period (Fig. 9-11). However, during a 2-hour incubation, the addition of increasing concentrations of LDL only showed a tendency to enhance PGE_2 production. In contrast to the lack of short-term effects of LDL on PGE_2 synthesis, the addition of oxidized LDL to mesangial cells resulted in a rapid and concentration-dependent increase in PGE_2 synthesis (Fig. 9-11). Prolonged (24 hours) incubation of LDL with mesangial cells results in oxidative modification of the LDL.[17] It therefore appears likely that the modest effect on PGE_2 synthesis during 24-hour incubations of mesangial cells with LDL may in fact be due to oxidized LDL generated during the incubation period. These findings are reminiscent of those reported for lipoprotein-induced eicosanoid synthesis in other cells where the oxidized form of LDL appears to mediate the effects on prostaglandin production.[124]

CONCLUSION

Under physiologic conditions, the amount of LDL in the interstitial space depends on the balance between LDL delivery and removal. These fluxes in turn are influenced by multiple factors. We shall schematically summarize these factors with special reference to the mesangium, but again point out that similar considerations apply to the renal interstitium as well.

The delivery of LDL to the mesangium is influenced by the following:
- the concentration of LDL in the plasma
- the net filtration forces in the glomerular capillaries
- the permeability of the endothelial–mesangial interface to LDL

The removal of LDL from the mesangium is dependent on the following:
- the binding and trapping of LDL to the mesangial matrix
- the uptake of LDL by mesangial cells and resident macrophages
- the efflux of LDL via the intramesangial channels

Under physiologic conditions, these factors will be in balance and LDL will provide the cholesterol necessary for mesangial cells. However, during disease processes, any of the above factors may be altered, leading to mesangial (interstitial) LDL accumulation. Accumulation of LDL may then initiate a

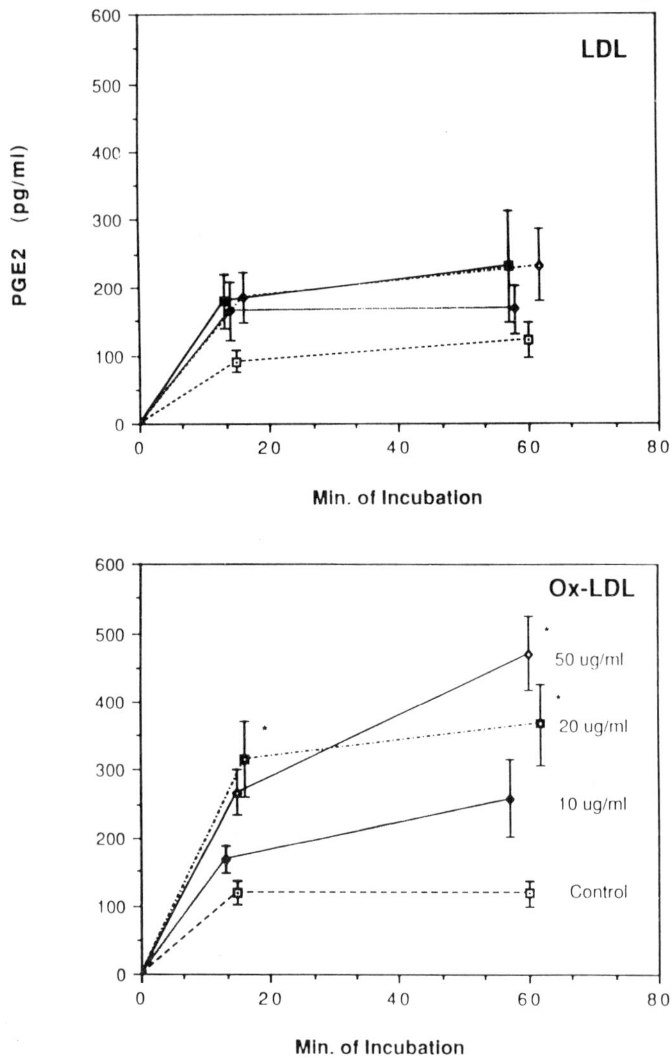

Fig. 9-11. Time course for PGE_2 synthesis following incubation of mesangial cells with variable concentrations of LDL or oxidatively modified LDL (Ox-LDL). Asterisks indicate $P < .05$ or better compared with the respective control values. (From Coritsides et al.,[34] with permission.)

pathologic cascade that, in analogy to atherosclerosis, eventually leads to mesangial and/or interstitial sclerosis.

A hypothetical scenario (Fig. 9-12) could be as follows. Renal diseases in association with hyperlipidemia and decreased anionic capillary charge may lead to enhanced mesangial delivery of LDL. The amount of LDL delivered to the mesangium would exceed the requirements of the mesangial cells, leading to a reduction in receptor-mediated LDL uptake and accumulation of

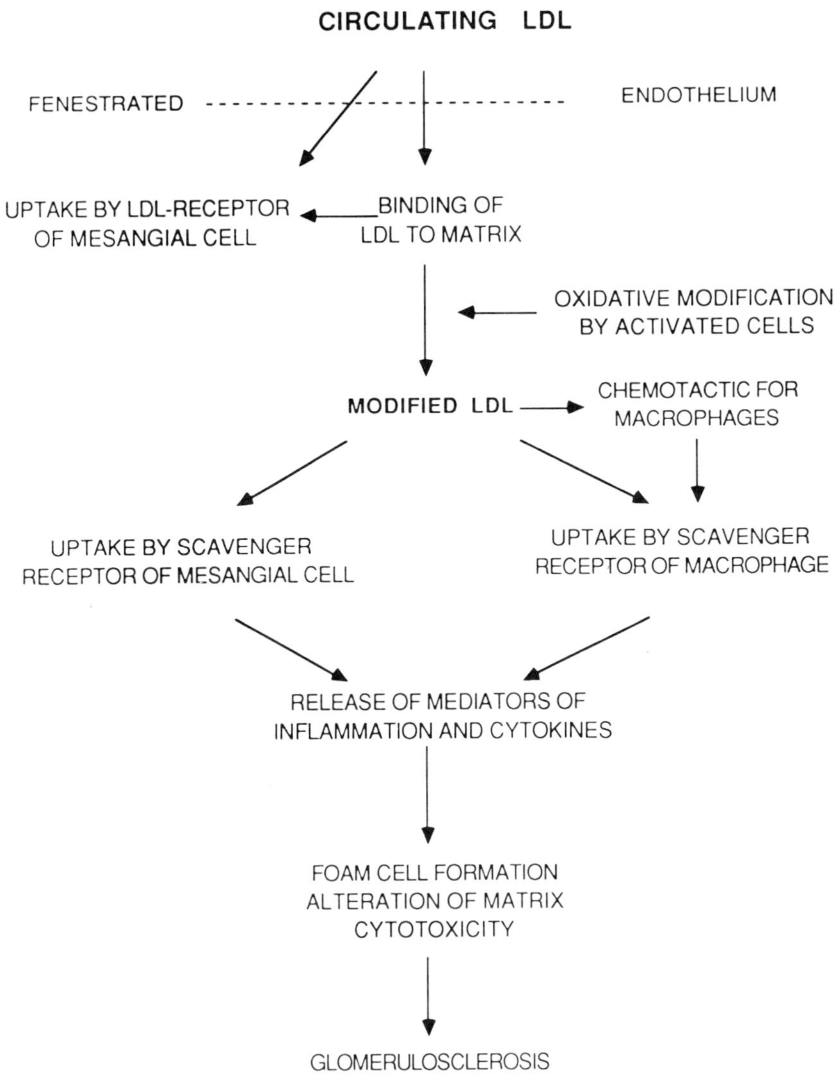

Fig. 9-12. Hypothetical scheme for metabolism of excess LDL accumulating in the glomerular mesangium.

LDL in the mesangium. Some of the excess LDL may leave the mesangium via intramesangial channels, but this would be insufficient to maintain balance. Thus, LDL would accumulate within and bind to the mesangial matrix.

In the mesangial microenvironment, LDL bound to matrix components such as collagen and proteoglycans would be subject to modification (e.g., oxidation by mesangial or endothelial cells or even resident macrophages). Modification of LDL in the mesangial microenvironment of the normal glomerulus would be limited by circulating inhibitors of oxidation. In contrast, inflammatory glomerular diseases, which are characterized by enhanced generation of reactive oxygen species by activated mesangial cells and/or infiltrating monocyte macrophages,[18] may lead to increased modification of LDL within the diseased mesangium. LDL could also be modified by forming complexes with glycosaminoglycan and other components of matrix. Modified LDL would be taken up by scavenger receptors on mesangial cells and monocyte macrophages. If the amount of modified LDL formed is moderate, a new steady state may be reached in which formation and removal of modified LDL remain in balance. This would result in the transformation of mesangial cells and macrophages to foam cells. Alterations in cytokine production and eicosanoid synthesis and release of urokinase and plasminogen activator inhibitors may ensue.[34,80] Modification of LDL may play an additional role in the accumulation of macrophages due to chemotactic effects and inhibition of macrophage egress.[129]

Should continued excess delivery of LDL to the mesangium exceed the newly established balance, more LDL would be trapped in the mesangium, undergo modification, and be taken up in part by scavenger receptors. Additional accumulation of oxidized LDL and foam cell formation may then result in cytotoxic effects, leading to death of mesangial cells, further activation of macrophages, and enhanced matrix formation. Oxidized LDL-induced stimulation of eicosanoid synthesis by macrophages and mesangial cells may in turn influence local hemodynamics,[46] vascular permeability,[130] matrix deposition,[79,131] and cytokine production.[80] The potential importance of enhanced eicosanoid synthesis to hemodynamic changes in hyperlipidemia is illustrated by recent studies in cholesterol-fed rats.[46] In these studies, hyperlipidemia induced renal vasoconstriction, which was largely mediated by increased thromboxane synthesis. In turn, enhanced thromboxane and PGE_2 production appeared to be related to oxidation of LDL, as they were prevented by probucol treatment. Increased thromboxane production may also contribute to the development of glomerular sclerosis by stimulating matrix production.[132] In addition, lipoproteins themselves may stimulate synthesis of proteoglycans and other matrix components. More matrix, in combination with activated macrophages, will continue a vicious cycle of LDL trapping, oxidation, and cell toxicity. Eventually, these mesangial changes will have detrimental effects on the adjacent epithelial and endothelial cells. Damage to the endothelial cells will expose the underlying thrombogenic matrix and cause thrombosis and obliteration of the glomerular capillaries. The endothelial damage with denudation of the capillaries would thus result in a catastrophic acceleration of the glomerular disease process, leading to either loss of capillary lobules or total glomerular loss with global sclerosis. Moreover, glomerular diseases are also characterized by marked interstitial changes and prominent interstitial foam cell accumulation. The association of glomerular diseases with interstitial changes may be in part linked by abnormal lipid accumulation.

It should be stressed again that this scheme requires the concurrence of at least two major factors: (1) hyperlipidemia and (2) alterations in the mesangial or interstitial microenvironment favoring local LDL modification. Inhibitors of LDL oxidation normally present in the glomerular microenvironment might be overwhelmed during inflammatory glomerular disease processes by enhanced generation of oxygen-free radicals, leading to enhanced local modification of LDL within the mesangium. Thus, hyperlipidemia per se would not be sufficient to cause glomerular or interstitial sclerosis, but would become an accelerating factor in the presence of local inflammation and injury. This hypothesis might help to explain why hyperlipidemia secondary to renal disease appears to contribute to the progressive deterioration of renal function, while primary hyperlipidemias rarely lead to glomerulosclerosis.

ACKNOWLEDGMENT

This work was supported in part by grants 2R01 DK 22036 and 5R01 DK41566. We gratefully acknowledge the expert secretarial help of Mrs. I. Baptiste.

REFERENCES

1. Venkatachalam MA, Karnovsky MJ: Extravascular protein in the kidney. Lab Invest 27:435, 1972
2. Kesaniemi YA, Witzum JL, Steinbreicher UP: Receptor-mediated catabolism of low density lipoprotein in man: quantitation using glucosylated low density lipoprotein. J Clin Invest 71:950, 1983
3. Carew TE, Pittman RC, Steinberg D: Tissues sites of degradation of native and reductively methylated [^{14}C] sucrose-labeled low density lipoprotein in rats. J Biol Chem 257:8001, 1982
4. Brown MS, Basu SK, Falck JR et al: The scavenger cell pathway for lipoprotein degradation: specificity of the binding site that mediates the uptake of negatively-charged LDL by macrophages. J Supramolecular Structure 13:67, 1980
5. Mahley RW, Innerarity TL, Pitas RE et al: Inhibition of lipoprotein binding to cell surface receptors of fibroblasts following selective modification of arginyl residues in arginine-rich and B apoproteins. J Biol Chem 252:7279, 1977
6. Brown MS, Goldstein JL: A receptor-mediated pathway for cholesterol homeostasis. Science 232:34, 1986
7. Goldstein JL, Ho YK, Basu SK, Brown MS: Binding site on macrophages that mediates uptake and degradation of acetylated low density lipoprotein, producing massive cholesterol deposition. Proc Natl Acad Sci USA 76:333, 1979
8. Steinbrecher UP: Oxidation of human low density lipoprotein results in derivatization of lysine residues of apolipoprotein B by lipid peroxide decomposition products. J Biol Chem 262:3603, 1987

9. Steinberg D, Parthasarathy S, Carew TE et al: Beyond cholesterol: modifications of low-density lipoprotein that increase its altherogenicity. N Engl J Med 320:915, 1989
10. Steinbrecher UP, Parthasarathy S, Leake DS et al: Modification of low density lipoprotein by endothelial cells involves lipid peroxidation and degradation of low density lipoprotein phospholipids. Proc Natl Acad Sci USA 81:3883, 1984
11. Brown MS, Goldstein JL: Scavenging for receptors. Nature 343:508, 1990
12. Kodama T, Freeman M, Rohrer L et al: Type 1 macrophage scavenger receptor contains alpha-helical and collagen-like coiled cells. Nature 343:531, 1990
13. Rohrer L, Freeman M, Kodama T et al: Coiled-coil fibrous domains mediate ligand binding by macrophage scavenger receptor type II. Nature 343:570, 1990
14. Matsumoto A, Naito M, Itakura H et al: Human macrophage scavenger receptors: primary structure, expression, and localization in atherosclerotic lesions. Proc Natl Acad Sci USA 87:9133, 1990
15. Via DP, Pons L, Dennison DK et al: Induction of acetyl-LDL receptor activity by phorbol ester in human monocyte cell line THP-1. J Lipid Res 30:1515, 1989
16. Morel DW, DiCorleto PE, Chisolm GM: Endothelial and smooth muscle cells alter low density lipoprotein *in vitro* by free radical oxidation. Arteriosclerosis 4:357, 1984
17. Keane WF, Phillips J, Kasiske BL et al: Injurious effects of low density lipoprotein (LDL) on human mesangial cells (HMC), abstracted. Kidney Int 37:509, 1990
18. Boyce NW, Tipping P, Holdsworth SR: Glomerular macrophages produce reactive oxygen species in experimental glomerulonephritis. Kidney Int 35:778, 1989
19. Heinecke JW, Rosen H, Chait A: Iron and copper promote modification of low density lipoprotein by human arterial smooth muscle cells in culture. J Clin Invest 74:1890, 1984
20. Steinbrecher UP: Role of superoxide in endothelial-cell modification of low-density lipoproteins. Biochim Biophys Acta 959:20, 1988
21. Morel DW, Hessler JR, Chisolm GM: Low density lipoprotein cytotoxicity induced by free radical peroxidation of lipid. J Lipid Res 24:1070, 1983
22. Sparrow CP, Parthasarathy S, Steinberg D: Enzymatic modification of low density lipoprotein by purified lipoxygenase plus phospholipase A_2 mimics cell-mediated oxidative modification. J Lipid Res 29:745, 1988
23. Parthasarathy S, Young SG, Witzum JL et al: Probucol inhibits oxidative modification of low density lipoprotein. J Clin Invest 77:641, 1986
24. Morel DW, and Chisolm GW: Antioxidant treatment of diabetic rats inhibits lipoprotein oxidation and cytotoxicity. J Lipid Res 30:1827, 1989
25. Hoff HF, Gaulatz JW: Isolation, purification, and characterization of a lipoprotein containing apo B from the human aorta. Atherosclerosis 42:273, 1982
26. Goldstein JL, Hoff HF, Ho YK et al: Stimulation of cholesterol ester synthesis in macrophages by extracts of atherosclerotic human aortas and complexes of albumin/cholesteryl esters. Atherosclerosis 1:210, 1981
27. Yla-Herttula S, Palinski W, Rosenfeld MD et al: Evidence for the presence of oxidatively modified low density lipoprotein in atherosclerotic lesions of rabbit and man. J Clin Invest 84:1086, 1989
28. Haberland ME, Fong D, Cheng L: Malondialdehyde-altered protein occurs in atheroma of Watanabe heritable hyperlipidemic rabbits. Science 241:215, 1988

29. Palinski W, Rosenfeld ME, Yla-Herttuala S et al: Low density lipoprotein undergoes oxidative modification *in vivo*. Proc Natl Acad Sci USA 86:1372, 1989
30. Fogelman AM, Schechter I, Seager J et al: Malondialdihyde alteration of low density lipoproteins leads to cholesterol ester accumulation in human monocyte-macrophages. Proc Natl Acad Sci USA 77:2214, 1980
31. Mahley RW, Innerarity TL, Weisgraber KH, Oh SY: Altered metabolism (*in vivo* and *in vitro*) of plasma lipoproteins after selective chemical modification of lysine residues of the apoproteins. J Clin Invest 64:743, 1979
32. Wasserman J, Santiago A, Rifici V et al: Interactions of low density lipoprotein with rat mesangial cells. Kidney Int 35:1168, 1989
33. Rayner HC, Horsburgh T, Brown SL et al: Receptor-mediated endocytosis of low-density lipoprotein by cultured human glomerular cells. Nephron 55:292, 1990
34. Coritsides G, Rifici V, Gupta S et al: Preferential uptake of oxidized LDL over LDL by cultured mesangial cells and by glomeruli *in vivo* and effects on mesangial cells. Kidney Int 39:858, 1991
35. Walli A, Grone HJ, Abboud HE et al: Characterization of receptor-specific uptake of apo B and E containing lipoproteins and their interaction with endothelia in human mesangial cells in culture, abstracted. J Am Soc Nephrol 1:734, 1990
36. Lee LY, Mohler WA, Schafer BL et al: Nucleotide sequence of the rat low density lipoprotein receptor cDNA. Nucleic Acids Res 17:1259, 1989
37. Grone HJ, Walli AK, Grone E et al: Receptor mediated uptake of apo B and apo E rich lipoproteins by human glomerular epithelial cells. Kidney Int 37:1449, 1990
38. Goldstein JL, Basu SK, Brunschede GY, Brown MS: Release of low density lipoprotein from its cell surface receptor by sulfated glycosaminoglycans. Cell 7:85, 1976
39. Witzum JL, Mahoney EM, Branks MJ et al: Nonenzymatic glucosylation of low-density lipoprotein alters its biologic activity. Diabetes 31:283, 1982
40. Schleicher E, Deufel T, Wieland OH: Non-enzymatic glycosylation of human serum lipoproteins. FEBS Lett 129:1, 1981
41. Lorenzi M, Cagliero E, Markey B et al: Interaction of human endothelial cells with elevated glucose concentrations and native and glycosylated low density lipoproteins. Diabetologia 26:218, 1984
42. Gonen B, Baenziger J, Schonfeld G et al: Nonenzymatic glycosylation of low density lipoproteins in vitro. Effects on cell interactive properties. Diabetes 30:875, 1981
43. Lopes-Virella MF, Klein RL, Lyons TJ et al: Glycosylation of low-density lipoprotein enhances cholesteryl ester synthesis in human monocyte-derived macrophages. Diabetes 37:550, 1988
44. Curtiss LK, Witztum JL: Plasma apoliproteins, A1, A2, B, C1 and E are glucosylated in hyperglycemic diabetic subjects. Diabetes 34:452, 1985
45. Schleicher E, Oglemoller B, Schon J et al: Limited nonenzymatic glucosylation of low-density lipoprotein does not alter its catabolism in tissue culture. Biochim Biophys Acta 846:226, 1985
46. Kaplan R, Aynedjian HS, Schlondorff D, Bank N: Renal vasoconstriction caused by short-term cholesterol feeding is corrected by thromboxane antagonist or probucol. J Clin Invest 86:1707, 1990
47. Pittman RC, Attie AD, Carew TE, Steinberg D: Tissue sites of degradation of

low density lipoprotein: application of a method for determining the fate of plasma proteins. Proc Natl Acad Sci 76:5345, 1979

48. Chen RM, Getz GS, Fischer-Dzoga K, Wissler RW: The role of hyperlipidemic serum on the proliferation and necrosis of aortic medial cells *in vitro*. Exp Mol Pathol 26:359, 1977
49. Ronnemaa T, Doherty NS: Effect of serum and liver extracts from hypercholesterolemic rats on the synthesis of collagen by isolated aortas and cultured aortic smooth muscle cells. Atherosclerosis 26:261, 1977
50. Daoud AS, Fritz KE, Jarmolych J: Increased DNA synthesis in aortic explants from swine fed a high-cholesterol diet. Exp Mol Pathol 13:377, 1970
51. Fischer-Dzoga K, Fraser R, Wissler RW: Stimulation of proliferation in stationary primary cultures of monkey and rabbit aortic smooth muscle cells. Exp Mol Pathol 24:346, 1976
52. Bourdillion MC, Boissel JP, Crouzet B: Proliferation of primary cultures from rat aortic media. Prog Biochem Pharmacol 14:103, 1977
53. Robertson AL, Jr: Role of circulating lipoproteins in the proliferative phase of atherogenesis. Am J Pathol 74:49a, 1974
54. Koschinsky T, Bunting CE, Rutter R, Gries FA: Increased growth stimulation of human vascular cells by serum from patients with primary hyper-LDL-cholesterolemia. Atherosclerosis 63:7, 1987
55. Fischer-Dzoga K, Wissler RW, Vesselinovitch D: The effect of estradiol on the proliferation of rabbit aortic medial tissue culture cells induced by hyperlipemic serum. Exp Mol Pathol 39:355, 1983
56. Fischer-Dzoga K, Wissler RW: Stimulation of proliferation in stationary primary cultures of monkey aortic smooth muscle cells. Part 2. Effect of varying concentration of hyperlipemic serum and low density lipoproteins of varying dietary fat origins. Atherosclerosis 24:515, 1976
57. Fischer-Dzoga K, Kuo Y-F, Wissler RW: The proliferative effect of platelets and hyperlipidemic serum on stationery primary cultures. Atherosclerosis 47:35, 1983
58. Ledet T, Fischer-Dzoga K, Wissler RW: Growth of rabbit aortic smooth-muscle cells cultured in media containing diabetic and hyperlipemic serum. Diabetes 25:207, 1976
59. Brown BG, Mahley R, Assmann G: Swine aortic smooth muscle in tissue culture. Circ Res 39:415, 1976
60. Tauber J-P, Cheng J, Gospodarowicz D: Effect of high and low density lipoproteins on proliferation of cultured bovine vascular endothelial cells. J Clin Invest 66:696, 1980
61. Fless GM, Kirchhausen T, Fischer-Dzoga K et al: Serum low density lipoproteins with mitogenic effect on cultured aortic smooth muscle cells. Atherosclerosis 41:171, 1982
62. Augustyn JM, Fritz KE, Daoud, AS, Jarmolych J: Effect of lipoprotein on *in vitro* synthesis of DNA in aortic tissue. Atherosclerosis 27:179, 1977
63. Traill KN, Huber LA, Wick G, Jurgens G: Lipoprotein interactions with T cells: an update. Immunol Today 11:411, 1990
64. Wosu L, Parisella R, Kalant N: Effect of low density lipoprotein on glycosaminoglycan secretion by cultured human smooth muscle cells and fibroblasts. Atherosclerosis 48:205, 1983
65. Koschinsky T, Bunting CE, Schwippert B, Gries FA: Increased growth of human fibroblasts and arterial smooth muscle cells from diabetic patients related to diabetic serum factors and cell origin. Atherosclerosis 33:245, 1979

66. Evensen SA, Sverre K, Nilsen E: LDL-induced cytotoxicity and its inhibition by anti-oxidant treatment in cultured human endothelial cells and fibroblasts. Atherosclerosis 49:23, 1983
67. Morel DW, Chissolm GM: Antioxidant treatment of diabetic rats inhibits lipoprotein oxidation and cytotoxicity. J Lipid Res 30:1827, 1989
68. Henriksen T, Evensen SA, Carlander B: Injury to cultured endothelial cells induced by low density lipoproteins: protection by high density lipoproteins. Scand J Clin Lab Invest 39:369, 1979
69. Chisolm GM, Morel DW: Lipoprotein oxidation and cytotoxicity: effect of probucol on streptozotocin-treated rats. Am J Cardiol 62:20B, 1988
70. Abrocast BW, Lee GM, Raymond TL: *In vitro* injury of porcine aortic endothelial cells by very-low density lipoproteins from diabetic rat serum. Diabetes 31:593, 1982
71. Hessler JR, Morel DW, Lewis J, Chisolm GM: Lipoprotein oxidation and lipoprotein-induced cytotoxicity. Arteriosclerosis 3:215, 1983
72. Henriksen T, Evensen SA, Carlander B: Injury to human endothelial cells in culture induced by low density lipoproteins. Scand J Clin Lab Invest 39:361, 1979
73. Hessler JR, Robertson AL, Jr, Chisolm GM: LDL-induced cytotoxicity and its inhibition by HDL in human vascular smooth muscle and endothelial cells in culture. Atherosclerosis 32:213, 1979
74. Rapoport J, Franki N, Hays RM: An autoradiographic study of urea transport: the effect of phloretin on the luminal entry of [^{14}C]urea. Trans Assoc Am Physicians 97:283, 1984
75. Chisolm GM, Hessler JR, Morel DW, Lewis LJ: *In vitro* autooxidation renders low density lipoprotein toxic to cells in culture, abstracted. Arteriosclerosis 1:359a, 1981
76. Hessler JR, Morel DW, Lewis LJ, Chisolm GM: Lipoprotein oxidation and lipoprotein-induced cytotoxicity. Arteriosclerosis 3:215, 1983
77. Mulcahy WS, Hetzel CA, Mills DCB et al: Lysophosphatidylcholine is the major cytotoxic component of oxidized LDL inproliferating rat mesangial cells, abstracted. J Am Soc Nephrol 1:636, 1990
78. Kosugi K, Morel DW, DiCorleto PE, Chisolm GM: Toxicity of oxidized low-density lipoprotein to cultured fibroblasts is selective for S phase of cell cycle. J Cell Physiol 130:311, 1987
79. Mene P, Simonson MS, Dunn MJ: Physiology of the mesangial cell. Physiol Rev 69:1347, 1989
80. Groene HJ, Abboud HE, Walli AK et al: Effect of low density and beta-very low density lipoproteins on PDGF mRNA expression and DNA synthesis in human mesangial cells, abstracted. Am Soc Nephrol 1:455, 1990
81. Hoff HF, Jackson RL, Mao SJ, Gotto AM, Jr: Localization of low-density lipoproteins in atherosclerotic lesions from human normolipemics employing a purified fluorescent-labeled antibody. Biochim Biophys Acta 351:407, 1974
82. Hoover GA, McCormick S, Kalant N: Interaction of native and cell-modified low density lipoprotein with collagen gel. Arteriosclerosis 8:525, 1988
83. Hollander W: Unified concept on the role of acid mucopolysaccharides and connective tissue proteins in the accumulation of lipids, lipoproteins, and calcium in the atherosclerosis plaque. Exp Mol Pathol 25:106, 1976
84. Steele RH, Wagner WD, Rowe HA, Edwards IJ: Artery wall derived proteoglycan-plasma lipoprotein interaction: lipoprotein binding properties of extracted proteoglycans. Atherosclerosis 65:51, 1987

85. Hoff HF, Wagner WD: Plasma low density lipoprotein accumulation in aortas of hypercholesterolemic swine correlates with modifications in aortic glycosaminoglycan composition. Atherosclerosis 61:231, 1986
86. Mawhinney TP, Augustyn JM, Fritz KE: Glycosaminoglycan-lipoprotein complexes from aortas of hypercholesterolemic rabbits. Atherosclerosis 31:155, 1978
87. Srinivason SR, Dolan P, Radhakrishnamurthy B, Berenson GS: Isolation of lipoprotein-acid mucopolysaccharide complexes from fatty streaks of human aortas. Atherosclerosis 16:95, 1972
88. Srinivasan SR, Dolan P, Radhakrishnamurthy B et al: Lipoprotein-acid mucopolysaccharide complexes of human atherosclerotic lesions. Biochim Biophys Acta 388:58, 1975
89. Iverius PH: The interaction between human plasma lipoproteins and connective tissue glycosaminoglycans. J Biol Chem 247:2607, 1972
90. Srinivasan SR, Lopez SA, Radhakrishnamurthy B, Berenson GS: Complexing of serum pre-beta and beta-lipoproteins and acid mucopolysaccharides. Atherosclerosis 12:321, 1970
91. Vijayagopal P, Srinivasan SR Radhakrishnamurthy B, Berenson GS: Interaction of serum lipoproteins and a proteoglycan from bovine aorta. J Biol Chem 256:8234, 1981
92. Camejo G, Waich S, Mateu L et al: Differences in the structure of plasma low-density lipoproteins and their relationship to the extent of interaction with arterial wall components. Ann N Y Acad Sci 275:153, 1976
93. Camejo G, Ponce E, Lopez F et al: Partial structure of the active moiety of a lipoprotein complexing proteoglycan from human aorta. Atherosclerosis 49:241, 1983
94. Camejo G, Lopez A: The very low density lipoproteins of cholesterol-fed rabbits: a study of their structure and *in vivo* changes in plasma. Atherosclerosis 19:139, 1974
95. Mateu L, Avila EM, Camejo G et al: The structural stability of low-density lipoprotein. A kinetic x-ray scattering study of its interaction with arterial proteoglycans. Biochim Biophys Acta 795:525, 1984
96. Bihari-Varga M, Sztatisz J, Gal S: Changes in the physical behavior of low density of lipoprotein in the presence of glycosaminoglycans and high density lipoprotein. Atherosclerosis 39:19, 1981
97. Yla-Herttuala S, Jaakkola O, Solakivi T, Kuivaniemi H: The effect of proteoglycans, collagen and lysyl oxidase on the metabolism of low density lipoprotein by macrophages. Atherosclerosis 62:73, 1986
98. Hurt E, Camejo G: Effect of arterial proteoglycans on the interaction of LDL with human monocyte-derived macrophages. Atherosclerosis 67:115, 1987
99. Vijayagopal P, Srinivasan SR, Jones KM et al: Complexes of low-density lipoproteins and arterial proteoglycan aggregates promote cholesteryl ester accumulation in mouse macrophages. Biochim Biophys Acta 837:251, 1985
100. Falcone DJ, Salisbury BGJ: Fibronectin stimulates macrophage uptake of low density lipoprotein-heparin-collagen complexes. Arteriosclerosis 8:263, 1988
101. Salisbury BGJ, Falcone DJ, Minick CR: Insoluble low-density lipoprotein-proteoglycan complexes enhance cholesteryl ester accumulation in macrophages. Am J Pathol 120:6, 1985
102. Basu SK, Brown MS, Ho YK, Goldstein JL: Degradation of low density lipoprotein. Dextran sulfate complexes associated with deposition of cholesteryl esters in mouse macrophages. J Biol Chem 254:7141, 1979

103. Goldstein JL, Hoff HF, Ho YK et al: Stimulation of cholesteryl ester synthesis in macrophages by extracts of atherosclerotic human aortas and complexes of albumin/cholesterol esters. Arteriosclerosis 1:210, 1981
104. Alavi MZ, Moore S: Proteoglycan composition of rabbit arterial wall under conditions of experimentally induced atherosclerosis. Atherosclerosis 63:65, 1987
105. Srinivasan SR, Radhakrishnamurthy B, Dalferes ER, Jr, Berenson GS: Collagenase-solubilized lipoprotein-glycosaminoglycan complexes of human aortic fibrous plaque lesions. Atherosclerosis 34:105, 1979
106. Smith FB, Massie IB, Alexander KM: The release of an immobilized lipoprotein fraction from atherosclerotic lesions by incubation with plasmin. Atherosclerosis 25:71, 1976
107. Brownlee M, Vlassara H, Cerami A: Nonenzymatic glycosylation products on collagen covalently trap low-density lipoprotein. Diabetes 34:938, 1985
108. Brownlee M, Vlassara H, Kooney A et al: Inhibition of diabetes-induced arterial wall lipoprotein deposition and prevention of arterial wall collagen crosslinking by aminoguanidine, abstracted. Diabetologia 29:523A, 1986
109. Yijayagopal P, Sprinivasan SR, Dalferes ER, Jr et al: Effect of low-density lipoproteins on the synthesis and secretion of proteoglycans by human endothelial cells in culture. Biochemistry 255:639, 1988
110. Ehrlich K, Murray M: Effect of low density lipoprotein on proteoglycan synthesis by aorta cells in culture. Experientia 34:179, 1978
111. Sudhakaran PR, Prinz R, Filipovic I et al: Homologous low density lipoprotein does not affect proteoglycan metabolism of cultured skin fibroblasts and arterial smooth muscle cells. Hoppe-Seyler Z Physiol Chem 361:129, 1980
112. Pietila K, Nikkari T: Effect of phase of growth and hyperlipidemic serum on the synthesis of collagen in rabbit aortic smooth muscle cells in culture. Med Biol 56:11, 1978
113. St. Clair RW, Jones DC, Hester SH: Failure of hypercholesterolemic serum to stimulate collagen synthesis in aortic smooth muscle cells from two species of nonhuman primates having different rates of collagen synthesis. Proc Soc Exp Biol Med 174:137, 1983
114. Jarvelainen H, Halme T, Lehtonen A, Ronnemaa T: Serum from type 2A hyperlipoproteinemic patients does not stimulate proliferation of and collagen synthesis in human fetal aortic smooth muscle cells in culture. Atherosclerosis 56:199, 1985
115. Holderbaum D, Ehrhart LA, McCullagh KG: Effects of hyperlipoproteinemic serum and exogenous proline concentration on collagen synthesis by isolated rabbit aortas. Proc Soc Exp Biol Med 150:363, 1975
116. Holderbaum D, Ehrhart LA: Inhibition of collagen and non-collagen protein synthesis in cultured aortic smooth muscle cells by hyperlipoproteinemic serum. Artery 7:16, 1980
117. Ronnemaa T, Jarvelainen H, Lehtonen A et al: Growth of human aortic smooth muscle cells cultured with human serum is retarded when serum lipids are lowered by medroxyprogesterone. Atherosclerosis 67:223, 1987
118. Orekhow AN, Tertov VV, Kudryashov SA, Smirnov VN: Triggerlike stimulation of cholesterol accumulation and DNA and extracellular matrix synthesis induced by atherogenic serum or low density lipoprotein in cultured cells. Circ Res 66:311, 1990
119. Crowley S, Brownlee M, Singhal P et al: Effects of nonenzymatic glycosylation

(NEG) of matrix on rat mesangial cell (MC) function, abstracted. Kidney Int 37:218, 1990
120. Pomerantz KB, Tall AR, Feinmark SJ, Cannon PJ: Stimulation of vascular smooth muscle cell prostacylin and prostaglandin E_2 synthesis by plasma high and low density lipoproteins. Circ Res 54:554, 1984
121. Fleisher LN, Tall AR, Witte LD et al: Stimulation of arterial endothelial cell prostacyclin synthesis by high density lipoproteins. J Biol Chem 257:6653, 1982
122. Szczeklik A, Gryglewski RJ: Low density lipoproteins (LDL) are carriers for lipid peroxides and inhibit prostacyclin (PGI_2) biosynthesis in arteries. Artery 7:488, 1980
123. Beitz J, Forster W: Influence of human low density and high density lipoprotein cholesterol on the *in vitro* prostaglandin I_2 synthetase activity. Biochim Biophys Acta 620:352, 1980
124. Yokode M, Kita T, Kikawa Y et al: Stimulated arachidonate metabolism during foam cell transformation of mouse peritoneal macrophages with oxidized low density lipoprotein. J Clin Invest 81:720, 1988
125. Stuart MJ, Gerrard JM, White JC: Effect of cholesterol on production of thromboxane B_2 by platelets *in vitro* N Engl J Med 302:6, 1980
126. Haga Y, Takata K, Araki N et al: Intracellular accumulation of cholesteryl esters suppresses production of lipopolysaccharide-induced interleukin 1 by rat peritoneal macrophages. Biochem Biophys Res Commun 160:874, 1989
127. Cushing SD Berliner JA, Valente AJ et al: Minimally modified low density lipoprotein induces monocyte chemotactic protein 1 in human endothelial cells and smooth muscle cells. Proc Natl Acad Sci USA 87:5134, 1990
128. Schlondorff D, Mori T: Contributions of mesangial cells to glomerular immune functions. Klin Wochensch 68:1138, 1990
129. Quinn MT, Parthasarathy S, Fong LG, Steinberg D: Oxidatively modified low density lipoproteins: a potential role in recruitment and retention of monocyte/macrophages during atherogenesis. Proc Natl Acad Sci USA 84:2995, 1987
130. Schlondorff D, Ardaillou R: Prostaglandins and other arachidonic acid metabolites in the kidney. Kidney Int 29:108, 1986
131. Schlondorff D: The glomerular mesangial cell: an expanding role for a specialized pericyte. FASEB J 1:272, 1987
132. Klotman P, Bruggeman L, Hassell J et al: Regulation of extracellular matrix by thromboxane, abstracted. Kidney Int 35:294, 1989

Index

Page numbers followed by f *indicate figures; those followed by* t *indicate tables.*